P9-DNL-482

River Forest Public Library
735 Lathrop Avenue
River Forest, IL 60305
708-366-5205
March 2019

SKELETON KEYS

ALSO BY BRIAN SWITEK

FOR ADULTS

Written in Stone

My Beloved Brontosaurus

FOR CHILDREN

Prehistoric Predators

SKELETON KEYS

The SECRET LIFE *of* BONE

BRIAN SWITEK

RIVERHEAD BOOKS NEW YORK 2019

RIVERHEAD BOOKS
An imprint of Penguin Random House LLC
penguinrandomhouse.com

Drawings in chapters 1, 4, 5, 6, 9, and 10: *Osteographia, or the Anatomy of Bones* by William Cheselden (London, 1733). From the Wellcome Collection.

Drawings in the introduction and chapters 7 and 8: *Osteographia, or the Anatomy of Bones* by William Cheselden (London, 1733). From the National Library of Medicine Collection.

Drawings in chapters 2 and 3: After William Cheselden, 1830/1835? From the Wellcome Collection.

Library of Congress Cataloging-in-Publication Data
Names: Switek, Brian, author.
Title: Skeleton keys : the secret life of bone / Brian Switek.
Description: New York : Riverhead Books, 2019. | Includes bibliographical references and index.
Identifiers: LCCN 2018037813 (print) | LCCN 2018039921 (ebook) |
ISBN 9780399184918 (ebook) | ISBN 9780399184901
Subjects: LCSH: Human skeleton. | Bones.
Classification: LCC QM101 (ebook) | LCC QM101 .S5487 2019 (print) | DDC 611/.71—dc23
LC record available at https://lccn.loc.gov/2018037813

Printed in the United States of America
10 9 8 7 6 5 4 3 2 1

BOOK DESIGN BY MEIGHAN CAVANAUGH

For Fox

So you can better trace
my skeletal stories.

CONTENTS

SKELETON KEYS

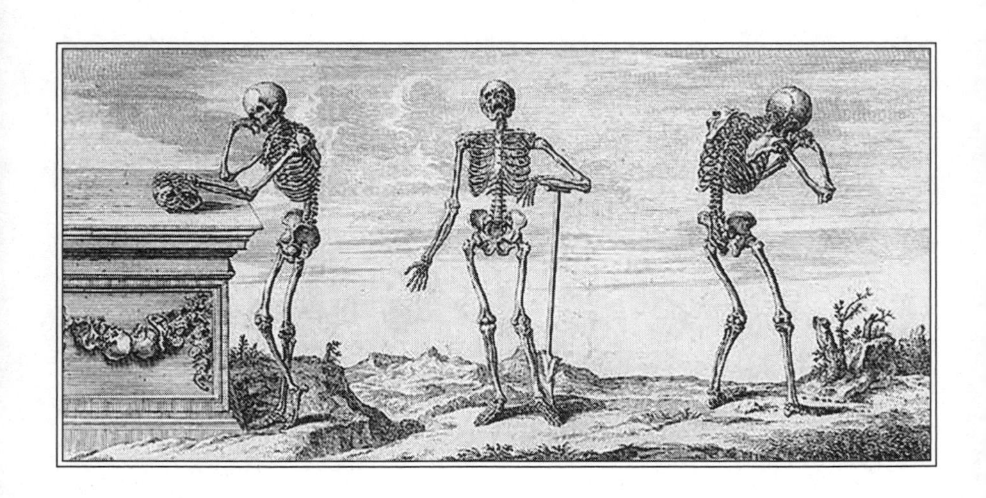

INTRODUCTION

CUT TO THE BONE

When Geza Uirmeny decided to take his own life, he turned to a blade. Precisely what was tormenting the seventy-year-old Eastern European shepherd is a secret kept by his remains. The tiny placard affixed beneath his toothless skull in a cabinet of wood and glass at downtown Philadelphia's Mütter Museum doesn't say whether it was financial stress, heartache, or any of the other painful circumstances of human life that led him to choose his own way out. But the postmortem grin formed by his jaws speaks to what happened next.

What Uirmeny didn't know when he raised the cutting edge to his neck was that part of his throat had transformed to bone. This happens to everyone to a greater or lesser extent. The flexible cartilage of your larynx—the ringed tube that gives you that distinctive part of yourself, your voice—slowly but surely starts to change as you age, rigid bone cells growing in place of the more pliant flesh.

Uirmeny's tissues were a little more ambitious than most. As he swiped the blade across his neck, he met unexpected resistance. His larynx had been so transmuted that it formed a bony strut in his neck; in the more clinical terms of the Mütter Museum's signage, "Wound not fatal because of ossified larynx." That little note doesn't record what Uirmeny felt when he realized his failure, but the scar that must have formed on his neck was a mark of a happier ending. Uirmeny, the display says, "lived until 80 without melancholy." Bone saved Uirmeny's life.

The lucky herdsman's skull is one of 139 in the *Hyrtl Skull Collection* exhibit, the last resting place for dozens of people who perished during the second half of the nineteenth century in Central and Eastern Europe. Each skull has its own story, recounted in a passive voice shorthand that makes the collected tales swing between the somber and the tragicomic. There's the bony grin of Francisca Seycora, a nineteen-year-old Viennese prostitute who died of meningitis, next to that of Veronica Huber, a woman executed for murdering her child. They share the space with rail workers, fishermen, bandits, soldiers, and the unidentified dead, as well as a few stranger cases. There's the cranium of Andrejew Sokoloff, a member of an extreme religious sect who died following the order's dire requirement of self-emasculation; and the skull of Girolamo Zini, a twenty-year-old tightrope walker who, the museum's deadpan delivery tells us, "died of atlanto-axial dislocation (broken neck)."

These crania aren't the only bones in the Mütter's expansive and historic collection. In addition to housing slices of Einstein's brain and larger-than-life replicas of every eye injury imaginable, which sympathy pain prevented me from giving any more than a sideways glance, the Mütter Museum is home to the towering skeleton of the Mütter American Giant, the remains of a woman so tightly corseted for so long that the garments changed the very structure of her bones, and dozens of other people whose final act is to educate the rest of us about what lives inside. This is a place populated by the remarkable dead, a medical mausoleum with a nineteenth-century aesthetic that would make a Victorian-era anatomy student feel right at home. There's more than a touch of the gothic about the rows of cases, not to mention a sinister feeling that you, too, might have been eyed for an exhibit had you lived during the museum's heyday—anatomists of old would often let their ethics slip if it meant acquiring another remarkable specimen for their cabinets. These bones have taken on a second life, and that's part of the story as much as the lives stripped down and presented among the cases and shelves. Every bone in the collection embodies a tangle of stories leading from the present back through history, and deeper still through the millions of years of evolutionary assembly that made us into our present and still-changing form. In each skeleton, whether that person was privileged or poor, healthy or afflicted, there are stories of varied and resilient *life*.

Admittedly, I hadn't paid very close attention to human bones before my visit to the Mütter that cold February morning. My affection for bones had its origin in paleontology.

I lived just an hour from the Mütter for most of my life, and I always promised myself I'd visit at an unspecified *sometime*. When I found myself with enough time and cash to visit a museum, though, I'd always opt instead to go north on NJ Transit to visit the hulking frames of dinosaurs and other prehistoric oddities in Manhattan's

American Museum of Natural History. Petrified bones of all shapes and sizes fascinated me, even more glorious raised and reconstructed in their life positions.

That enduring affection led me to settle in the American West, where I spend weeks out of each summer helping museum and university field crews dig up fragile bones that throw open windows to lost worlds. It's difficult work. Out in the desert, science is a matter of stomping around crumbling outcrops in search of pieces of prehistoric lives that have miraculously survived to the present, using pick, shovel, brush, and plaster to uncover and cradle old bones before using whatever strength you can muster to drag them out of their natural tombs. All that manual labor offers plenty of time to think, of course, and the endless stream of questions that bones inspire helps those wracked with fossil fever endure sunburn, gnat bites, dehydration, and cactus needles that seem to know the exact weak spot of your boots.

What was this creature? What did it look like? How did it move? What did it eat? These are puzzles that can be answered through bone. Each element has stories to tell, a record of the organism's life wrapped up in its skeleton. To the paleontologist, bones are not grim visages of death. Skeletons are biological time capsules that tell us of lives we'll never see in the flesh. A tooth. A string of vertebrae. An osteoderm that once acted as bony armor inside the skin. These were all living tissues that had to grow and were constantly changing within the bodies of the animals they once belonged to. Even the tiniest, most boring fragment of unidentifiable Chunkasaurus gradually turning to powder beneath the unforgiving sun is a vestige of a life come, gone, and preserved for a span of time that's impossible to truly understand. It's difficult to push away thoughts about life while facing death. This is as true for us as it was for *Tyrannosaurus*.

Gently, insistently interrogating the remains of long-dead creatures makes every tidbit of information drawn from their skeletons a treasure. We don't get to see them in life, so bones are most of what we have. (Tracks and traces form a supplement to the skeletons.) The entire paleontological discipline is based on resurrecting the extinct, if only in our minds.

With our own bones, though, the connection flips. We intimately experience life and are familiar with all the squishy tissues that skeletons support. With the knowledge of the living, then, the meaning of human bones is often pulled inside out. A skull is a death's-head, reminding us of what awaits us all. "As I am now, so will you be. As you are now, so once was I." That's what human skeletal remains repeat to us over and over again. Just think of where we see skeletons and skulls around us. A skull and crossbones marks the menacing flutter of the Jolly Roger. A similar symbol warns us we'll die if we're careless about what containers we drink from. Heavy metal album covers are rife with skeletons, as are the ranks of fictional villains, from Skeletor to the bony army in *The 7th Voyage of Sinbad*. Tattooed onto my left forearm, a werewolf grips the skull of one of her victims in an anthropomorphic memento mori. Even Death itself comes to us as a robed skeleton. One of the few positive cultural associations with skeletons we seem to have is the Mexican Dia de los Muertos, a holiday when sugar skulls and other osteological adornments help keep the living in touch with the memory of those who have left us. But that's largely an exception to our modern relationship with bones. While prehistoric remains represent life resurrected and reassembled, we often think of our own bones as potent symbols of the afterlife and what misfortune may come to snatch us away into it.

Still, deep down, we can't escape the fact that bones are the most vital of structures. They form the foundation of our lives from the outset. We don't squeeze out of our mothers fully formed, all of our 206-some-odd bones in place. No, when we start life our bones still have to fully ossify and fuse, years and years of development and growth ahead of us that will alter our skeletons as the years tick by. Our bones change throughout our childhood and adolescence, eventually bringing us to our full stature, and if we're lucky enough to reach old age, they can even turn against us, resorbing themselves before we're finished using them. These changes don't happen in fits and starts. Bone is always being altered. Even now, as you're reading this, specialized, voracious cells are eating up old bone as other cellular blobs are creating new bone cells, recycling your body from the inside. So even though there's a useful distinction between flesh and bone, it really only comes down to soft versus hard. Bone is every bit as lively and responsive as the skin that hides all the muscle and gore draped around your skeleton.

Bone, in scientific jargon, is "a vascularized tissue consisting of osteocytes with multiple interconnecting processes, embedded in an extracellular matrix, mineralized with hydroxyapatite, and containing type I collagen." Like many scientific definitions, this manages to be correct and totally miss the larger point at the same time. Yes, bone is a durable, mineralized tissue made of a hard part and a flexible part, but it's also one of the most fantastic building materials that evolution has accidentally spit out. In us bone is a structural core, giving our bodies support while also acting as a foundation for our flesh and internal protection that wraps around our vital organs. It never moves by itself but is essential to our ability to move. Bone has manifested itself in wings, sails, horns, armor, and an even greater array of appendages and ornaments that have been accumulating since the time of its origin. Bone is the vital stuff that made possible something as small as the 0.6 inch-long Jaragua lizard—as tiny as a raindrop—as well as the 110-foot-long dinosaur *Supersaurus* and the 190-ton blue whale, the largest animal to have yet evolved. Stomp, fly, swim, slither, dig, run—all are expressions of what bones make possible. But that's hardly all.

Science has a habit of reducing the complex and nuanced to the narrow and then working back up to the broader picture. That's part of how we construct a way of thinking about the natural world. We need constant, agreed-upon definitions in order to organize and understand. But winnowing down an observation or idea to its purest and most minimalistic form isn't always wise. John Steinbeck pointed this out long before I did. Reflecting on his time floating along the Sea of Cortez with legendary marine biologist Ed Ricketts, Steinbeck wrote that a scientist could easily look at a fish like a Mexican sierra and boil that animal down to measurements and notes on spine count. But that wouldn't really capture the whole truth of the animal. Both the quantifiable aspects of the animal and the poetic, elusive nature of its existence are needed. "We determined to go doubly open so that in the end we could, if we wished, describe the sierra thus: 'D. XVII-15-IX, A. II-15-IX,' but also we could see the fish alive and swimming, feel it plunge against the lines, drag it threshing over the trail, and even finally eat it." So it is with us, and our bones. We can define bone by its biochemical components, its evolutionary history, its shape and variation, but to boil down bone to numbers, measurements, or landmarks will always seem incomplete. It doesn't matter whether we're talking about a skeleton donated to a museum, a skull stolen from a grave, or even a bone chip left weathering to nothing out on the ground. The truth of the bone very much depends on who's looking at it, and our skeletons are as embedded in our culture as they are in our bodies. Our species has made instruments and jewelry from bone, treated the dead like collector's items, put our faith in skull bumps as guides to human behavior, arranged skeletons into macabre tributes to the afterlife, and more.

While bone itself can tell our individual stories, we've also woven our changing relationship with our intimate osteology into our wider history. (Osteology being the study of bones. If you see the prefix *osteo*, you know you're dealing with something involving skeletons.) To borrow a term from paleontology, there is a stratigraphy of stories surrounding bone—layer upon layer of detail that tells us who we are and where we came from. All it takes is asking the right questions.

That's where my own journey started. Dinosaurs and saber-toothed cats initially drew my attention to bone—how could they not?—but I wanted to let my curiosity approach humanity in the same fashion as it did my favorite petrified monsters. I had mostly ignored human bones in my research and writing—stripped down, we just can't compete with the grandeur of something like *Stegosaurus*—but my ignorance kept itching me. If I didn't know human skeletons, I didn't really know myself. I wanted to remedy that, and other skeletons had already given me a blueprint for doing so. After all, we're all built on the same osteological chassis with many of the same parts, inherited from a common fishy ancestor that binds vertebrates together into one great family. It follows that the same approaches to understanding other forms of vertebrate life can be turned onto us, too, allowing us to tease apart the layers of what our bones tell us about who we are, where we came from, and the constant identity crisis that comes part and parcel with being human. After all, we are not separate from nature but an especially unusual manifestation of it. I wanted to take the same approach to the bones in my body as I would the burly frame of a mastodon or the threatening bulk of an *Allosaurus*. Perhaps in doing so, I could begin to understand where I fit in among life's fantastic skeletal menagerie I so admire. That drive led me to the stories I'm about to share with you.

What I want to do, in essence, is cut down to the bone. I have a deep affection for that osteological expression. You can be dry as a bone, be reduced to a bag of bones, pick a bone of contention, and, in exasperation, ask to be thrown a bone, for starters. But there's something more visceral about cutting down to the bone, slicing through to the underlying truth of something. The flesh can, and often does, lie. It's easily malleable stuff. Untruths spring from our squishy brains and chicanery is carried out by the push and pull of all those soft tissues. But bone remains pure, forming the very core of us. Perhaps it's the idea of dissecting away everything else, reducing ourselves to our struts and supports. What better example of hard fact than bone itself?

Not that the denuded truth is easy to look at. Coming face-to-face with what's inside you can be a haunting experience. A skeleton is a final and everlasting snapshot of concluded vitality, reminding us of our bodies' eventual conclusion. Skulls bring us face-to-face with this fate. Peering into glass museum cases to look into empty sockets that once held expressive eyes, I can't help but flinch. In those unsettling and tense moments, we remember that we are not just viewing objects—these were, and still are, people. And that's hardly all. Cutting through biology, history, and culture to understand the meaning of what's inside us isn't always a pleasant experience. As we get closer to the bone, there may be some parts of ourselves we don't wish to see, and others that challenge our preconceptions of who we are. Much of what we used to so fervently believe about bones has turned out to be wrong. Reliquaries holding the remains of the holy have become the stuff of myth and legend, especially given how many of them do not contain the genuine remains they're supposed to preserve. We no longer look to skulls for signs of whether or not someone has a criminal disposition. And, contrary to what some anthropologists of generations past so fervently believed, vast collections of skulls accumulated in the name of science have shown that race has no biological meaning, creating a spectrum in place of the finely divided categories craniometrists were so certain existed.

It was never the bone itself that led us astray. Bald facts mean nothing without interpretation, and those hypotheses and analyses and arguments tell us just as much about ourselves as the bones themselves.

In the course of writing this book, I've inadvertently made myself a speaker for the dead, though I've tried to turn the narratives over to the original owners of those bones as much as possible. That's the responsibility of telling stories, even as the result ends up a tale that's as unique as my own skeleton. Bone is alive, wondrously and vibrantly so. Every element holds clues about the lives we lead. Human bones are not inert objects or curios, but were once part of people who had their own ideas, values, and beliefs. We intersect with them more powerfully than any osteological part of any other species. This is a wellspring of both wonder and terror, the pivot point bones occupy in our lives and culture.

Naturally, I felt most comfortable starting in the distant prehistoric past, long before human skeletons—or even bone itself—existed. We need to briskly stroll through the hundreds of millions of years in which our skeletons were assembled to understand how we came to be as we are. It's a humbling experience to look back in time and realize that there's no inevitability to our existence, especially when for so much of our history we've tried to convince ourselves that we were specially made for this world. From there we'll slide into the biology of our skeletons—the nature of our bones, what they can and cannot tell us when stripped of flesh. This leads us to the marvelous abilities of our skeletal system—how our complement of bones work together to let us move, and how they respond to injuries and afflictions imposed from the inside and out. This is the life of our bones. But human skeletons have historically had incredibly active afterlives. Human bones have been objects of worship, conquest, and curiosity for longer than we'll ever really know. They've been used to both honor and denigrate the dead, from the origins of religion through the genesis of science and up through this very day. And considering how our bones have the potential to be the most lasting parts of us, we'll leave the story considering how bones are the legacies we leave behind. They are the only parts of us that have a chance of telling others about our lives millions of years after all our earthly works have turned to dust, when we're prehistoric creatures ourselves, as dead as the dinosaurs.

Bones are totems of enduring life. They testify to more than five hundred million years of evolutionary history, documenting resilience and survival against the odds. They grow and change, acting as osteological recorders as they do so. And through our interpretations of what they mean, they reveal the best and worst of us. Within us are more stories than we will ever fully realize. Now, listen.

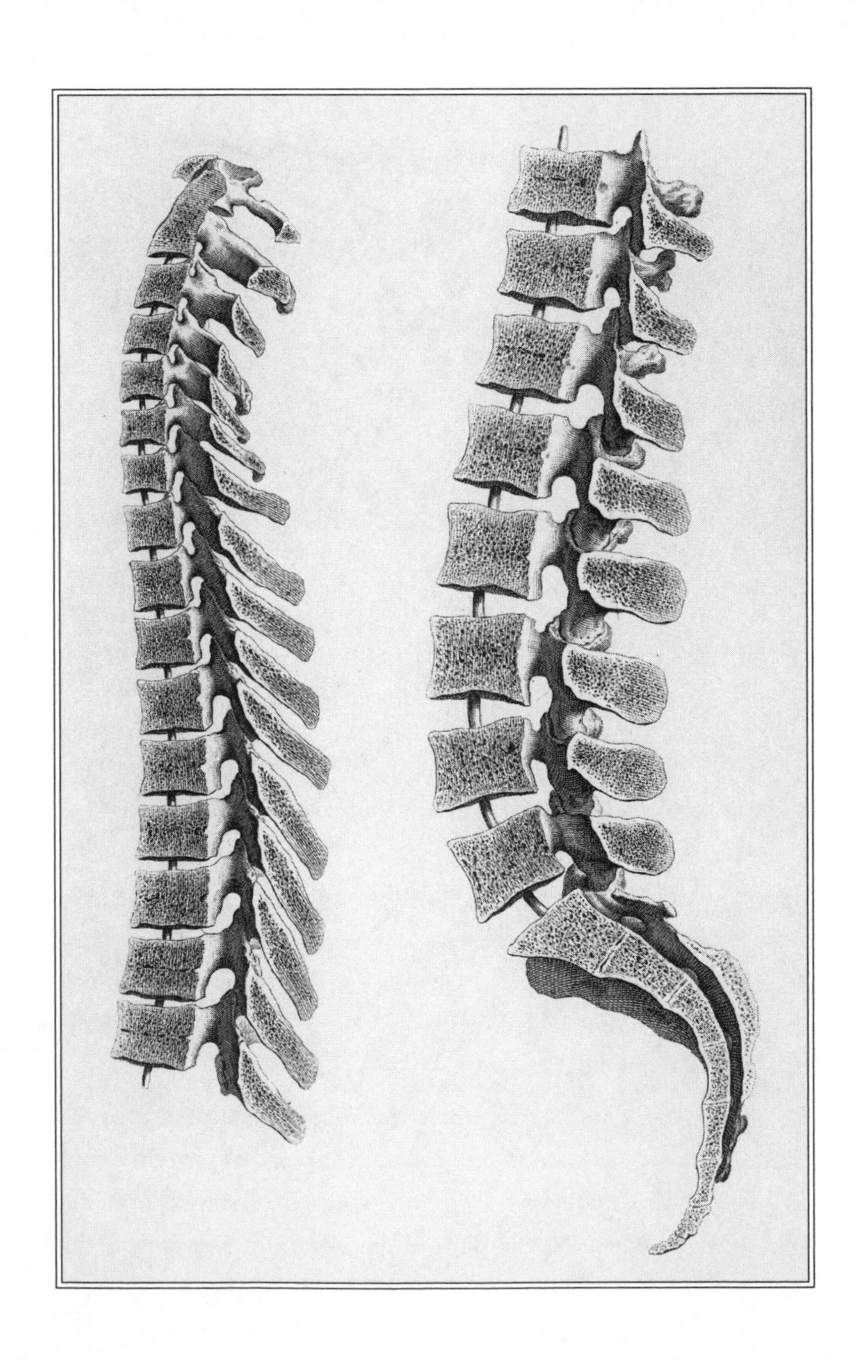

ONE

FLESH OUT

On a cool April afternoon, with two of my friends trailing behind me, I climbed the stone stairs of the National Museum of Natural History to meet Grover Krantz. A trip to the museum wasn't complete without a visit, and I had promised to introduce my companions to the famed and controversial anthropologist. (He published widely in the field, on everything from fossil apes to the evolution of human culture, but he is perhaps most remembered for his insistence that the mythical Bigfoot is real.) The appointment had been easy enough to make. That's because Krantz had been dead for more than a decade.

No special permit or permission was needed. Encased in glass, the skeleton of Krantz stood motionless at the end of the temporary *Written in Bone* exhibit, which was focused on the forensics of Chesapeake Bay residents in the days of colonization. Krantz was the most recent thing in the anthropological story, more of a coda than a part of it, and he didn't exactly fit with the European transplants arrayed in the osteological maze the Smithsonian had constructed. Still, he would have been happy for his inclusion at the end of the skeletal labyrinth. Krantz had made a career out of studying bones and teaching others their secrets. His dying wish was to continue that calling.

In his old age, after he was diagnosed with cancer, Krantz decided that internment in a grave or reduction to ashes wouldn't be fitting for someone who'd spent decades teaching anatomy. He wanted his bones to speak for him long after he died, and so he started pulling professional strings to find the perfect place to spend his afterlife. Krantz hoped that his gleaming bones would be reassembled for display alongside his three beloved Irish wolfhounds—Clyde, Icky, and Yahoo—whose bones he had been saving as the pups passed away. Macabre, maybe, but humans have had a long tradition of wanting to spend eternity with the ones we love.

The Smithsonian ultimately agreed to take Krantz, though public display seemed like a long shot. There was no place for him in the existing osteology hall, and the museum already had enough old skeletons hanging out in aging alcoves. Krantz would certainly have a home in the anthropological drawers, collections manager David R. Hunt promised, but Krantz's great rearticulation seemed like a fantasy when the anthropologist passed away from pancreatic cancer on Valentine's Day in 2002. His remains were sent to the famed Body Farm, the University of Tennessee's Anthropology Research Facility, and, denuded of flesh, arrived at the Smithsonian in 2003.

The anthropologist's bones could have stayed in a drawer with his dogs. There are worse fates than that. But the idea for *Written in*

Bone offered the opportunity to revive Krantz—or at least reassemble him—as a showstopper right at the end. So Smithsonian taxidermist Paul Rhymer was tasked with the job of fitting every little divot to its corresponding notch, creating a slightly altered X-ray version of a photo depicting a happy Clyde jumping to greet Krantz. (Rhymer had to change the positioning from that depicted in the photo to avoid giving the impression that Clyde was leaping to rip out Krantz's throat.) The two formed an excellent study in the vertebrate form. Almost all the bones in Clyde had their correspondence in Krantz, two expressions of more than five hundred million years of vertebrate evolution.

Leaning back to accept Clyde's canine embrace, Krantz acted as an emissary for the living bone in all of us. Asking any one person to represent humanity is a foolhardy task that contradicts the diversity we treasure, of course, but it's still helpful to see someone else's bones if you want to get a feel for what grows inside you. Reduced to osteology, Krantz's outside now represents your insides. And much like the deceased anthropologist, you have 206 bones, more or less. Without peeling away all the surrounding tissues, either after death or by way of a high-resolution full-body CT scan that would expose you to an unsafe level of radiation, there's no way to get a totally precise count. Leaving aside differences in the way we're born, accidents that might remove an appendage, and surgeries that replace our real bones with flexible facsimiles, even complete human skeletons vary in bone count. Each set of clattering bones is as individual as our personalities.

There are a few things I can say for sure about your skeleton, though. Your skull is balanced atop your neck. The stacked column of vertebrae runs along your back rather than along your front. All the necessary equipment for you to see, hear, smell, and taste the world is ensconced in your head rather than being distributed in different places across your body like some kind of Guillermo del Toro monster. None of these characteristics make us unquestionably human. The hallmarks of humanity are far more subtle, and mostly noticeable because every other human species that flourished over the past six million years perished and left us with a gap between us and the great apes. It's easier to make hard categorical divisions when extinction's removed your ancestors and cousins. But we can leave aside the subtleties that distinguish us as *Homo sapiens* for the moment. The basic format of where our arms and legs fit onto our bodies, the placement of our spine, the internal hug of our rib cage around our vital organs—all of these characteristics are widely shared with other vertebrate animals, from chimpanzees to swallows to *Triceratops* and the little red eft, so tiny you can scarcely believe that something so small could possibly have bones. As different as you are from a crocodile or a tuna or a house cat, your skeleton is laid out along the same body plan, and that's because we're all relatives descended from creatures that just happened to have such a conformation. They lived during a time before jaws, before spines, before even bone itself. You can find one of them at the Smithsonian several floors below where Krantz's skeleton stood.

You'd think that a species so integral to our skeletal layout would be enshrined at the center of every great paleontology hall in the world. What's left of them would rest on a velvet pillow lit from above, with visitors allowed into the darkened tomb alone or in pairs to spend a few moments with the animals to whom we owe the very core of our being. If star treatment is fitting for gems like the Hope diamond, surely part of our own deep past deserves as much respect. At the very least they ought to be given a prominent place, front and center, as an introduction to the great fossil galleries, their meek little forms giving us the essential context for everything that comes after. But there are no such honors for this particular relative of ours. The animals most important to the organization of our bodies just can't compete with the fossil attractions that draw the crowds. That's why the dinosaurs are usually tucked away somewhere that requires you to journey through other halls first, so that you might learn something in your dash to stand in the shadows of the great reptiles as our own furball ancestors did for more than 180 million years. The dinosaurs and the other fan favorite exhibits are the museum equivalents of summer blockbusters. Whether our enjoyment of them is high-minded and self-aware or just slack-jawed ecstasy, they're the attractions that get asses in the seats (or sore feet on marble floors). That makes the animals I'm about to introduce to you the equivalent of independent films—critically praised but lacking the grand spectacle.

Our critical character is tucked away in a quiet place, one where almost no one goes. As you walk up to the National Museum of Natural History's Sant Ocean Hall, take a left at the right whale dangling overhead to follow the hanging parade of fossil cetaceans back toward a little side room. This is the place where the museum's less popular fossil stars have aggregated themselves—the buglike trilobites, the coil-shelled ammonites, stalked crinoids bristling with spikes, and various other invertebrates that provide the intellectual backbone for how extinction has repeatedly stripped the planet's biota and how life has always sprung back. Here you'll find the animal I'm talking about, a fossilized doodle in a row of oddities. It's called *Pikaia gracilens*, and its relationship to us wasn't always so clear.

Pikaia, as well as the smattering of fossils displayed around it, all came from a spot in British Columbia whose name is known to every paleontologist, whether they've worked on the material that has spilled from the site or not: the Burgess Shale. The standard story of the site's discovery, as it was told to me by a paleontology professor years ago, goes like this: The field season of 1909 was coming to a close. Charles Doolittle Walcott, who had been scouring the ancient shales near the small town of Field for signs of early life, had come up almost entirely empty-handed. His plans for a momentous discovery dashed upon the very stones that were supposed to yield their secrets, he and his wife packed up their camp and started their way down the mountain as the first snowflakes of the season began to fall, confirming that the year's work really had come to an end. But then Mrs. Walcott's horse slipped on the chilled slabs of cracked stone, turning over a piece of rock that could have been easily overlooked. Charles spotted something strange on the chip the horse's hooves had flipped up. Impressed into the ancient sediment was a prehistoric crustacean unlike any the paleontologist had seen before. If not the beginning of the paleontological superstition, then it became the most cherished example—you will always make your best find on the last hour of the last day of fieldwork. That initial clue was all Walcott needed to return the next year and start prying up an entire menagerie of animals that he literally could not make heads or tails of.

This Paleozoic tale is the closest paleontologists have to a eureka moment in our canon of stories, and I can certainly understand the appeal. No matter how much you prepare, no matter how skilled an eye you have for the first glimmerings of fossils peeking out of the stone, you may be totally hosed if luck isn't with you. But as poetic paleontologist Stephen Jay Gould wrote in his book on the Burgess Shale, *Wonderful Life*, the classic Walcott story is fable, not fact. Walcott kept records of almost every day of his field season, including his first fossil finds. He made them on August 30 or 31 of 1909, Gould pointed out, with no sign of inclement weather. And Walcott's reaction was not wide-eyed wonderment; he simply noted that he found some "interesting fossils." That's all. And it's more fitting with how major discoveries often take shape. Big finds often start small and uncertain, usually with nothing more than a few curious bone fragments or some enigmatic smears on fine-grained stone. That's precisely what happened with Walcott. The day after his initial find, he located an even better spot, this one boasting three invertebrates completely new to science, and he collected a few more beautiful slabs and specimens before he and the rest of the party packed out on a warm, sunny September day.

The scientific fate of those specimens was just as circuitous. Walcott returned to the Burgess Shale for a second season in 1910, finding even more, but in the end he described only a small fraction of the massive fossiliferous haul he brought back to the Smithsonian. And in trying to understand these organisms, Walcott took what has been half-jokingly called the shoehorn approach, fitting the confounding jumbles of legs and spines and body segments into already-known groups of organisms. The Cambrian seemed to be a time of jellyfish, sponges, and shrimp, much more ancient but—with the exception of extinct groups such as trilobites—not especially different from the reefs that encrust the bottom of some seabeds today. Walcott's characterization of the fauna stayed in place for decades. But when paleontologists began to return to the Burgess Shale fossils in the 1960s with new ideas as well as new techniques to prepare the fossils in high relief, they found a community far stranger than Walcott could have ever imagined in even his most fevered fossil dreams.

Exhibited at the Smithsonian, Walcott's creatures present a gallery of alien body shapes, spindly legs, and googly eyes. Animals were still a novelty in the Cambrian, and it would be tens of millions of years before anything worthy of the title *giant* would evolve in Earth's oceans. Burgess Shale species seem disproportionately huge in depictions of this time because there's no scale to compare them to and they're too otherworldly for us to understand how all the pulsing, flapping, squirming parts relate to each other. They look like prehistoric equivalents of the shape-shifting monster in John Carpenter's *The Thing*, although at much smaller scale. Most of them you could hold in the palm of your hand or on the tip of your finger. Among the smallest of all is little *Pikaia gracilens*, just about an inch and a half in length. Of all the animals in the Burgess Shale, this is the wee beastie that we have kinship with.

Superficially, *Pikaia* may be the nadir of petrified impressiveness. The fossil looks like little more than a charcoal scribble on gray stone, shorter than the last few words in this sentence. And here, a word of caution is needed. At 530 million years old, *Pikaia* lived too far back in time for us to *directly* link the protovertebrate to ourselves through an unbroken line of descent. Paleontologists are adamant about such caveats. We can be confident that all life is connected in a great family tree that draws back to a single common ancestor—which may or may not have been the first life on Earth—but, to borrow the analogy the geologists Charles Lyell and Charles Darwin employed, we're missing a great many characters, words, sentences, and paragraphs from Life's great story. In the less than two hundred years paleontology has existed as a field of study, we've only just begun to brush the dust off all the geo-literary fragments that are out there, much less assemble them in their true order. Broad outlines are clear, but the details of family affiliations from parents to offspring species are almost always in contention, and the further back in time we go, the more concentration is needed to pick out who's who in the fossil record. That's why paleontologists often speak of transitional fossils, or species with transitional features—those species that help bridge what might seem to be different lineages, like the way the feathered *Archaeopteryx* connects nonavian dinosaurs with birds or how the mammal *Pakicetus* helps demonstrate the changes whales underwent as they went from land-lubbers to beasts of the sea. Such creatures are part of evolutionary moments that have gained our special attention, often involving major shifts in anatomy and natural history. From the stone pages of the planet, then, some narratives are already clear, and certain archetypes—if not ancestors—have emerged as heroes in the story. *Pikaia* is one such champion.

Pikaia was one of the earliest Burgess Shale species that Walcott named, and he introduced the world to the squashed fossil in a 1911 paper given the unprepossessing title "Middle Cambrian annelids." His entire treatment of the species lasts five short paragraphs, less than a page in length. Walcott saw little *Pikaia* as an annelid worm not altogether different from the night crawlers that squirm their way up onto the surface of a lawn when rain floods their burrows. "This was one of the active, free-swimming annelids that suggest the Nephthydidae of the Polychaeta," he wrote, which translates to a plain sandworm to you and me. But when paleontologist Simon Conway Morris later looked at the literal handful of *Pikaia* fossils that had been found, he did not see a worm. The tiny, almost imperceptible segments running down the two-inch animal's body were not the stacked bands of an annelid worm, but primitive myomeres—packages of fibers arranged in V shapes that are the forerunners of the skeletal muscles in our bodies. Diminutive *Pikaia* also had a distinct head end, marked by a pair of strange tentacles, but the most stunning revelation of all was preserved as a thin line of Paleozoic sheen along its back. *Pikaia* had the beginnings of the spine that, given the circumstances of another five hundred million years of transmutation, would come to hold our backs erect. The rod was not surrounded by bone; that stuff would not come into existence until more than a hundred million years later. But, as Conway Morris and his adviser reported in 1979, *Pikaia* had a notochord—the basic stiffened structure that would form the foundation of the backbone.

I had to smush my nose against the glass to get my nearsighted eyes close enough to *Pikaia* to see some of these details when I last visited this old friend of mine. But yes, there they were. And how amazing is that? There are relatively few dinosaurs, fossil mammals, and other creatures that even approach this level of preservation. Fossilization favors the sturdy. *Pikaia* was barely a wisp in the

Cambrian seas, yet conditions were just right to entomb a small portion of these creatures in sediment so fine that we know not only the shape their bodies took in life, but the intricate inner workings that connect them to us across deep time. In the wider view of evolution, drawing back from the tiny twig to get a sense of where that small part fits, *Pikaia* was one of the earliest chordates, the family to which we also belong.

Not that *Pikaia* was the only early chordate on the scene. As if the Burgess Shale were not stupendous enough, China has its own equivalent, known as the Chengjiang fauna. These rocks, between 520 and 525 million years old, boast their own fossil treasures, including three small cousins of *Pikaia*. The first two to be discovered, *Haikouichthys* and *Myllokunmingia* (try saying that ten times fast), were announced in 1999, with *Zhongjianichthys* following in 2003, and all look like simplified versions of the guppies I used to bring home from the pet store. They didn't have the little tentacles of *Pikaia*, which remain an unexplained oddity, but they still had the V-shaped muscles and bullet-shaped body plan, which probably were not noticed by the invertebrates that slurped them up in those early Cambrian seas.

What makes these protovertebrates so special is easier to see with our gift of hindsight. First off, *Pikaia* and other early chordates had a head. Not terribly shocking, I know, but it's still essential to how our bodies came to be the way they are. Your eyes, mouth, and nose, harboring some of your most critical senses, are all bound up in your head, in proximity to your brain. Had the anatomical state of things been otherwise for the early chordates, the sensory centers of vertebrates may have been placed on their rumps, or distributed over their bodies and requiring the evolution of quick nerve networks to communicate through all the disparate parts. But more important than that, the notochord that formed along the back of these species set the basic framework for what would become the backbone and all the parts attached to it. Sharks, emus, tree frogs, wildebeest, and, of course, *you* owe the basic construction of the skeleton to animals so close to invertebrates that the first of their kind ever found was mistaken for a worm.

The story of *Pikaia* is essential to everything that follows. These creatures, or creatures like them, established the basis for *why* we are how we are. *Pikaia* and its increasing array of relatives, through pure accident, ended up establishing what it is to be a vertebrate. But the point I want to underscore, articulated by Gould decades ago, is that there was nothing special or remarkable about our ancestors back in the Cambrian. In the context of life at the time, the eventual rise of the vertebrates is an underdog tale.

During the time of all those spectacular Burgess Shale species, most of the wonderful forms swimming and skittering about were species far distant from our own predecessors. If you were able to travel back in time to the Cambrian, and had the forethought to bring a submersible with you, the glorious dawn of vertebrate life would seem rather dimmed.

You bob along at the surface, concentrating to hold back the seasickness, carrying out one last check of the instruments before making your descent. Sweat is already starting to dampen your clothes—how could you have forgotten that in this time the Burgess Shale is just south of the equator?—but in a few more moments you close the hatch and start your journey to the bottom. And even though the opening dirge from the *Jaws* theme starts to play in the back of your mind as you descend to the reef below, you remind yourself that there's nothing to worry about. Not only are you cozy in your capsule, but it's over a hundred million years before the dawn of sharks. The most fearsome creatures you're going to see are only interested in snaffling up worms and other little morsels.

Impatient to start checking species off your list, you scan your field guide of the 150-some-odd species found in this place. Some of them, such as the bacteria, you're not going to see as anything more than specks in the water column or mats on the seafloor. But at the very least you know you'll be the only person to ever see a living trilobite—more than 33 percent of the Burgess Shale fossils found in your own era are of these arthropods and their relatives, members of the segmented invertebrates that include everything from grasshoppers to tarantulas and lobsters of the modern world. And after a few moments, you reach a Seussian seascape of strange tubes. Some grow in branching clusters, like underwater saguaro cactus, while others look like hairy cucumbers that have been cut in half. These are early sponges, the chief architects of the Cambrian reef. As your eyes adjust to this nest of pillowy porifera, the otherworldly shapes of the more charismatic animals begin to register in your vision. A priapulid—or "penis worm" that looks like a spiky version of exactly what its common name implies—darts back into its burrow. A trilobite disturbed by the movement in the water rolls up on itself in self-defense, looking like a stouter version of the roly-polys you used to find under the woodpile back home. A *Wiwaxia* is either too confident or too unaware to be bothered by such disturbances. The living pincushion pushes its spiky self over the muddy bottom in search of who knows what, passing by a wormlike critter walking on tube feet with sharp spines waving in the current—a species so strange that it was named *Hallucigenia.* And for a moment you pull the submersible up short as something that looks like an alien mothership speeds across the bow, undulating oarlike fins along the side of its body. That must be *Anomalocaris* on the prowl for something soft to stuff into its shutter-shaped mouth.

You'd likely run out of air and have to return to the surface, and hopefully your own time, before spotting a *Pikaia.* (And that's a good thing, for, as time-travel movies remind us, you wouldn't want to accidentally kill anything related to your ancestry.) Paleontologists who have pored over the thousands and thousands of Burgess Shale fossils at the Smithsonian, the Royal Ontario Museum, and the Geological Survey of Canada have found that early chordates comprised only 2 percent of the Burgess Shale fauna. Arthropods, sponges, algae, and worms were far more abundant than our relatives. Of course, the fact that *Pikaia* was a soft little thing without a bone in its body brought a slight bias against it being preserved into the fossil record, but regular mudslides that repeatedly buried the reefs caused these wonderful and alien creatures of the Burgess Shale to be extensively and delicately pressed into the rock. These happenstances, while disastrous for the Cambrian creatures, are why we can look back at this time with such extraordinary clarity. The mudslides buried the reef species so quickly and completely that even the most fragile creatures, such as our cousin *Pikaia*, were preserved. Bad luck for them, but good luck for us. It's as close as we can get to a Polaroid of these ancient ecosystems, and they make one particular paleontological point especially clear: the Cambrian was very much an invertebrate world, with weirdos far more eye-catching and numerous than *Pikaia*. If you were able to visit these ancient reefs without any understanding of how the following 508 million years would play out, you'd dismiss *Pikaia* as a boring little thread that was clearly taking the lazy route while invertebrates were experimenting with every possible body plan their carapaces could morph into.

For seventeen million years after their origin in those Cambrian seas, vertebrate ancestors were rare, marginal animals that had to evade the piercing, crushing mouthparts of their ravenous neighbors. And they were even luckier and more persistent than we previously realized. Back when Gould was writing *Wonderful Life*, there seemed to be a hard and fast break at the end of the Cambrian—one of the bright, harsh lines in the fossil record that marks a mass extinction. In divvying up the rocks of the world, pioneering geologists often unintentionally identified these catastrophes by noting the drastic differences on either side of a particular layer of stone. After the Cambrian, the bulk of the ludicrous species spread throughout the ancient seas seemed to totally vanish. There were no more living pincushions, nozzle-nosed foragers, boomerang-headed arthropods, or shutter-mouthed monstrosities. A more anatomically subdued collection of animals such as early trilobites and brachiopods made it through, with our own chordate ancestors joining the throng of lucky survivors. This was one of the most critical moments in evolutionary history for our species, Gould wrote, marking a time when life could have gone in a very different direction. If *Anomalocaris* or *Wiwaxia* had been spared—if the extinction had been canceled—then the evolutionary routes between *Pikaia* and the first true vertebrates would have been closed, totally erasing us from history and vastly altering the evolutionary course of what followed. Or so the story went. New finds have rewritten the tale.

Gould's underlying point still holds: what exists now relies on innumerable past events that opened particular evolutionary options while closing or constraining others. This is called contingency, and it's the same as when you wonder what would have happened to your life if you had screwed up the courage to ask your first crush on a date back in middle school, if you had taken that gap year in college like you wanted, or if you had avoided that questionable gas station burrito for lunch—just on a larger scale. The most striking example is when an asteroid smacked the planet sixty-six million years ago to spark a mass extinction that booted the dinosaurs from their dominance and eradicated various other forms of life, shaking things up enough to give mammals a chance at proliferating. The entire history of life on Earth pivoted on this single moment, and if it hadn't happened, the world would probably still be ruled by toothy, feathered saurians of every shape and size. Almost every slice of deep time has critical moments like this—some of which we can detect, others that are too small to sense.

There was no evidence of any asteroid impact, massive volcano belching greenhouse gases into the atmosphere, or other stark extinction trigger at the end of the Cambrian. The leading explanation for what happened was more a matter of Darwinian calculus. Paleontologists thought that the Cambrian menagerie simply lost the race for life, supplanted by novel species that had an edge over these more archaic forms. Newer, sleeker, more advanced organisms beat them into permanent submission. This was a concept that went all the way back to Darwin's *On the Origin of Species*—that descendant species will be better adapted to the prevailing conditions than the species that spawned them, and so the new will usually overtake the old. Paleontologists referred to this particular changing of the guard as the Ordovician radiation. It's when the seas started to take on a more modern appearance. Snails grazed among gardens of coral while the early cousins of sea stars, squid, and clams mixed with some Cambrian leftovers that managed to hold on, such as the trilobites.

But now we know the closing chapter of the Cambrian does not record a vast extermination of the weird, as was once thought. In 2010, a multi-institution team of paleontologists announced that they had described a vibrant community of Burgess Shale creatures found in rocks 443 to 485 million years old, long after the close of the Cambrian. The "extinction" at the end of the Cambrian wasn't because the animals disappeared. It was because the proper sort of sedimentology to preserve soft-bodied organisms became much rarer and suitable examples had not yet been found. In fact, the newly discovered deposit seemed to show something of a mix in which the ancient met the then-modern—there were the snails and nautiloids and crinoids of the Ordovician, yes, but there were also reef-building sponges, worms, soft-shelled arthropods, armor-covered animals similar to *Wiwaxia*, and enormous relatives of the truly bizarre *Anomalocaris*. No one had expected to find an entire community of animals that looked like they were straight out of the Burgess Shale mixed in with the species that were supposed to have utterly obliterated the competition. The old had survived alongside the new.

So what does this have to do with our little chordate friends? Well, in *Wonderful Life*, Gould argues that without a mass extinction of weird Cambrian species, our protovertebrate ancestors would have needed to adapt and adjust in different ways. This would have vastly altered evolutionary history in ways that would make the origin of *Homo sapiens* unlikely, if not impossible. Yet we now know that the end-Cambrian extinction really has been canceled, which makes our ancestry all the more peculiar. The protovertebrates continued to eke out an existence in an invertebrate-dominated world, twitching their little myomeres to escape the rapacious appetites of their neighbors. The ripples of contingency can still be felt. Had the early chordates gone extinct—or if they had their notochord along their belly, say, or led with their tails rather than their heads—evolutionary history would have been wildly different, more so than we're probably capable of imagining. But thanks to the fortuitous survival of so many Cambrian creatures, our backstory has just gotten that much deeper. To say that the likes of *Pikaia* were simply lucky now reads as an insult. They were survivors in their own right. Something else, something not yet fully understood, was going on as life flowered in the ocean realm, the dawn of the chordates taking a slower rise against the busy world of the invertebrates. And it's against that background that another chance event opened up even greater possibilities. The world was about to see bone for the first time.

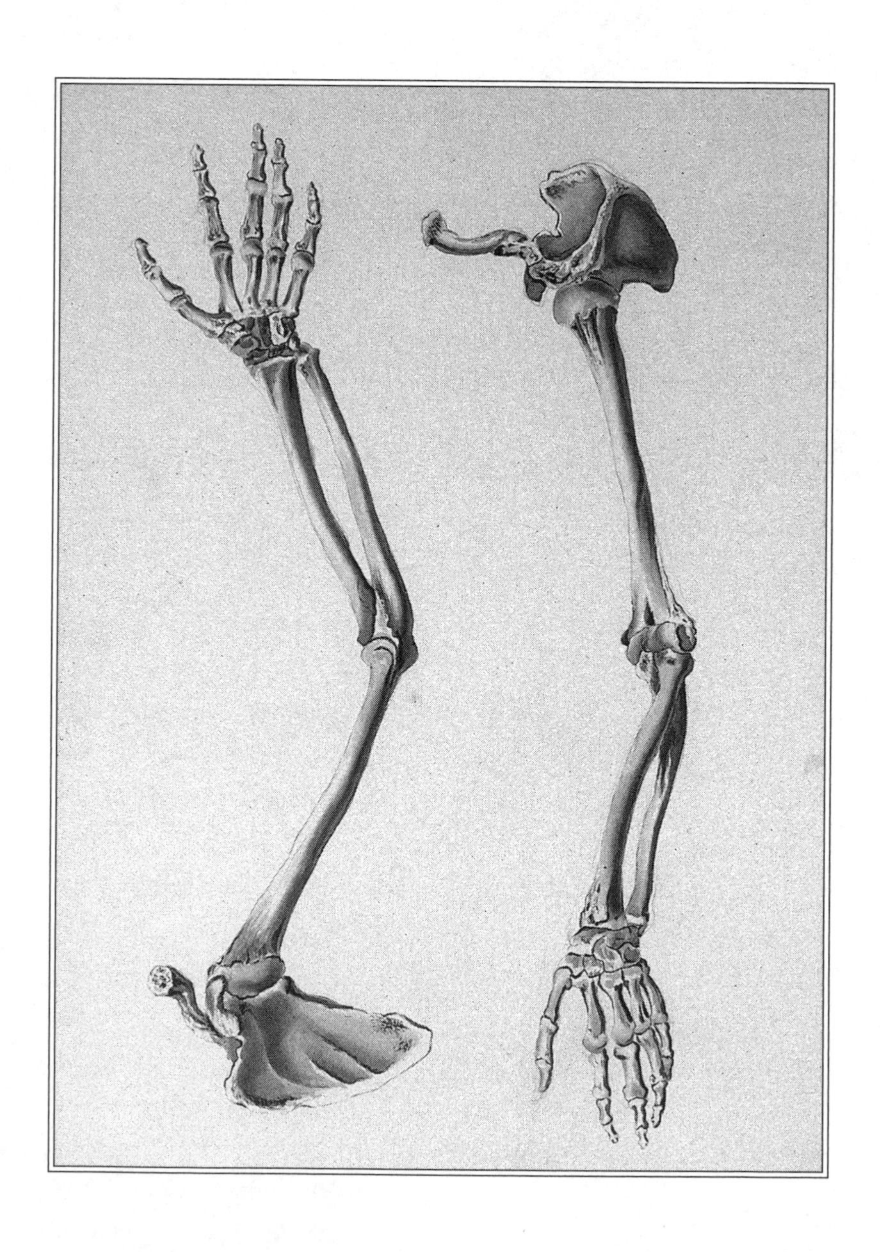

TWO

BONES TO PICK

Every science has its celebrities. Physics has Einstein. Chemistry has Marie Curie. Evolution, Charles Darwin. And in paleontology, it's two men who so thoroughly despised each other that they'd certainly be rankled by the fact that you can't mention the name of one without the other—Othniel Charles Marsh and Edward Drinker Cope.

The everlasting enmity between the two fossil hunters took time to develop. In fact, they started as friends. Both were young American scientists determined to be the late nineteenth century's paleontological pioneers. The discipline had not yet put down academic roots in the United States, and the two met while trying to glean what they could from the German authorities on the subject before establishing themselves in East Coast academic circles. They left behind almost no record of these early encounters—among the few amicable ones in their careers—but they got on well enough that, in 1868, Cope invited Marsh to visit one of the southern New Jersey marl pits that were keeping him well supplied with prehistoric fossils.

It was a sweet arrangement. The greensand was loaded with a mineral called glauconite, often used as a fertilizer, and the miners working the site regularly ran into the remains of creatures that had lived along the New Jersey coast more than sixty-six million years ago. Some, such as turtles and crocodiles, were vaguely familiar, but there were also tidbits of dinosaur and the fossils of gigantic seagoing lizards called mosasaurs. Even better for a naturalist with entire fossilized worlds waiting to be analyzed and described, the affluent Cope didn't have to dig through the muck himself. All he had to do was pay the miners for the better selection of specimens they encountered as they scraped the pits. Marsh—an ambitious man from a rich family that included his philanthropic uncle, George Peabody, who founded a museum at Yale for his nephew in 1866—could easily afford to do the same. When he saw the scientific potential of the fossils coming from Cope's connections, Marsh paid off the miners to send the bones north to his collection in New Haven rather than west to Cope's study in Philadelphia.

This was the paleontological equivalent of the shot heard round the world. Cope was furious that he had been undermined by Marsh, and this insult was just the first of many they'd exchange over the years. Their brilliance and arrogance made them the most cantankerous of adversaries. They practically dragged the entire field of paleontology into their squabble as they rushed to telegraph in species descriptions from western outposts and used their personal fortunes to fuel their fierce publication rate, always desperate to outdo each other.

The effects weren't all bad. The world was introduced to Mesozoic celebrities such as *Triceratops*, *Brontosaurus*, and *Ceratosaurus* as a result of this competition, which also changed the nature of the field. Not only were Cope and Marsh some of the first paleontologists to develop expertise in a wide variety of animals—describing fish, reptiles, amphibians, birds, and mammals as they pleased instead of focusing on one group—they also hired and trained those who would become the next generation of American paleontologists. Still, both Cope and Marsh were capable of holding a grudge with such endurance that in 1873 the editors of the scientific journal *The American Naturalist*—which Cope himself had purchased as a personal publishing platform—eventually refused to accept any more papers feeding the contest, informing readers that "the controversy between the authors in question has come to be a personal one and [because] the *Naturalist* is not called upon to devote further space to its consideration, the continuance of the subject will be allowed only in the form of an appendix at the expense of the author." That did nothing to quell the clash between the scientists. In 1890 the *New York Herald* brought the conflict public under the headline SCIENTISTS WAGE BITTER WARFARE, making other paleontologists feel like their entire profession had been publicly tarred. In the end, both men lost. Decades of trying to outdo each other had winnowed their personal finances to nearly nothing, and the stress had wracked their bodies. In 1897 as Cope lay dying in his personal museum, surrounded by his fossils and reptilian pets, he wasn't about to admit defeat to Marsh. Cope had plans to carry the Bone Wars into the afterlife.

No one knows exactly what killed Cope. There is no evidence to support the rumors that he succumbed to syphilis from a dalliance earlier in his life. He suffered for years from chronic infections and problems with his bladder, prostate, and neighboring tissues, for which he mostly medicated himself. Despite the urging of his friends, surgery to relieve some of his pains was out of the question. Such procedures were still in their infancy, historian Jane P. Davidson wrote in her biography of the scientist, and the prospect of accidentally being rendered impotent was one that Cope wasn't able to endure. He just continued to take belladonna—a poisonous plant that may have contributed to his death—and write up scientific papers when he could, locked in competition with Marsh until his final days. And he had a plan. Before he died on April 12, 1897, Cope set out a biological challenge for his Yale rival that would settle the question of who was the greatest paleontologist once and for all.

Cope did not care much for the bulk of his flesh. His muscles and nearly all of his vital organs were burnt to ash and placed in Philadelphia's Wistar Institute. But he had other wishes for the rest of his remains. "I direct that after my funeral my body shall be presented to the Anthropometric Society and that an autopsy shall be performed on it," Cope wrote in his will, stipulating that "My brain shall be preserved in their collection of brains" for future study. This was his final act of throwing down the gauntlet. He had given his brain to science to be weighed and measured, confident that the gray matter between his ears would outweigh that of Marsh. But Cope's nemesis never took the bait. Electing a burial less primitive than the fossils he loved, in 1899 Marsh was interred in a cemetery not far from the museum he had given his life to.

Contrary to rumors that it was lost or stolen, Cope's brain still rests in its liquid tomb in the Wistar Institute. And the rest of the master naturalist, as his student H. F. Osborn called him, isn't far away. In addition to his brain, Cope also left his skeleton to science, although he was not to be put on display. Cope did not have the afterlife showmanship of Grover Krantz, befitting the scientific spirit of the time. Dinosaurs and other fossil creatures were for the eyes of scientists only, not the public, and the sole dinosaur available for public view in the Americas was a reconstruction of the duck-billed *Hadrosaurus* erected at the Academy of Natural Sciences in 1868. So Cope's cleaned bones were kept under lock and key for the education of anthropology students.

As with almost everything else Cope-related, speculation and rumor cling to his bones. It has sometimes been said that Cope's real skull was dropped and smashed into irreparable pieces decades ago, but his cranium remains intact. In fact, it even had a bit of an unexpected and unauthorized adventure when, a century after Cope's death, photographer Louie Psihoyos and friend John Knoebber temporarily absconded with his cranium to bring the pioneer paleontologist face-to-face with his intellectual descendants from Generation X for their book, *Hunting Dinosaurs*. Even more widespread is the tale that Cope wanted his skeleton to become the holotype—or representative specimen—for *Homo sapiens*. That's easier to believe given the naturalist's oversize ego, but it's just a legend, too. It's also rather strange to think of any one skeleton holding the title of representative of our species. A holotype is the biological standard for all other animals of the same species to be compared against, but given the variety of human bodies, pointing to any one skeleton as the epitome of humanity is immediately problematic. Think about this the next time you're at the grocery store or a movie. Look around and consider any single one of those people becoming *the* representative of our species—it'd be an individual expression that'd obscure the wider diversity of who we are. No single person can represent the whole of our humanity.

Much like Krantz, however, Cope is still a useful starting point for the journey we're about to take. Just like you, he carried the familiar complement of 206-some-odd bones that has marked our species since we first evolved around three hundred thousand years ago. Every human skeleton—including yours—retains a history that goes back further than we can truly comprehend, part of a continuum of life in which the old is constantly being modified by increments into the new. The very arrangement of our skeletons is a mosaic woven through evolutionary time, from the origin of bone to the present day. There's too much to cover for this to be anything more than the flip-book version of the story, but we'll examine a few of the species that helped lay out how bones formed, came into articulation, were lost, and became modified into the form we associate with *Homo sapiens* today. If I'm at all successful, you'll be able to look at your body and see the hallmarks of ancient, fleshy-finned fish, snuffling little protomammals, the first primates, and other relatives of ours that present their internal structure to us in museum halls.

When dealing with such sweeping and consequential changes, context is everything. So let's rewind to start at the beginning.

The big bang birthed our universe about 13.8 billion years ago.

The happenstances of this event helped form the base elements that bone—and everything else—would eventually be made of. Then, as far as our tale is concerned, there was a nine-billion-year lull. About 4.54 billion years ago the beginnings of our planet accreted from vast clouds of cosmic dust and gas. Another four hundred million years on, the early Earth was given a cosmic smack by the protoplanet Theia to form the planet beneath our feet and the moon in our orbit. And even by this point, bone was still a long way off. The inscrutable spark that catalyzed life as we know it didn't happen until about 3.7 billion years ago, and this first organism had no hard parts to enhance its chances of entering the fossil record. We only know these dawn species existed at all because they immediately began changing the world, some of them pumping out oxygen in their efforts to sustain themselves, staining the rock with bands of rust as their ancient signature. And for a long, long time, life was content to stay in such a state: incapable of even aspiring beyond blankets of algae. Continents moved, climates swung from hot to cold to hot and back again, and life swarmed over the planet, making ours the Planet of the Microbes. (If we go by the numbers, it still is.) There were major changes during this time, of course. Cells subsumed other cells to create even more complex life, and in some forms DNA became wrapped up in a single nucleus instead of wafting unbounded within the cell. But for a vertebrate-focused fossil fan such as myself, everything was pretty humdrum until about half a billion years ago, when the Cambrian explosion that included *Pikaia*, *Anomalocaris*, and the other charismatic Burgess Shale creatures initiated the flowering of an ever-widening variety of fantastic species that has continued to this exact moment.

Even in the days of *Pikaia*, bone was at best a distant possibility. Life could have just as easily done without it if conditions had been different. The precursor of bone didn't appear until about 455 million years ago, thirty million years after the close of the Cambrian, and bone of the sort that acts as the framework inside our bodies didn't manifest itself until about 419 million years ago. While we can put our bony fingers on these points in time and say, "Aha! This is where our story begins," the origins of these tissues were not at all momentous. The armored fish that were the first to bear bone didn't cheer for the innovation or all immediately demand that their twitching little bodies be covered in this latest and greatest of building materials. If the armored fish of the day were able to discuss anything at all—made all the more impossible by their lack of jaws—they might have considered bone a passing fad. Bone was just an evolutionary happenstance that had the good fortune to be incredibly resilient, flexible, and useful in forming an amazing menagerie through the depths of time.

Bone is what made us and many of the animals we admire possible. It's the essential framework of many fish in the sea and creatures that have returned to the water, and bone was a deciding factor in the emergence of our ancestors onto land. Without bone, not only would we never have evolved, but neither would any of our favorite prehistoric oddities that required a strong support to literally stand against gravity. An internal, flexible bony skeleton is what allowed dinosaurs to range in size from a bee hummingbird, less than two inches long and practically as light as air at less than half an ounce, to 122-foot-long, seventy-ton titans far larger than anything that walked the earth before or since. Mammoths, giant sloths, and saber-toothed cats are all based in bone, too, as well as any other charismatic prehistoric vertebrate you care to name. And without the anatomical splendor of the variety of shapes and sizes that bone has allowed, the terrestrial realm may have been ceded to arthropods and other invertebrates. Life on Earth would have been incredibly different and unfolded on a much smaller scale. Contrary to Hollywood schlock, terrestrial invertebrates never would have attained the same monstrous sizes vertebrates eventually grew up to. There were giant dragonflies with two-foot-wide wingspans and six-foot-long millipedes at one point, back when the atmosphere's oxygen content was much higher than it is today, but an exoskeleton just cannot attain the enormous sizes that endoskeletons allow. Keeping all those guts in, rather than supporting them from the inside, becomes ever-more difficult with increasing size. An ant the size of the pinching hordes in *Them!* or a weevil the size of a VW Bug would immediately burst at its seams and crumple into a heap.

Bone, as it resides inside us, opened possibilities that would have otherwise gone unrealized. While they may not give daily thanks and praise, I'm certain paleontologists are grateful. Bones provide us with the most precious records we have of how life has changed through the ages. Just consider sharks. While sharks have skeletons, they're actually made of a more flexible material called cartilage (the stuff that supports the end of your nose and the fleshy expansions of your ears). This means that the fossil record of sharks primarily tracks teeth through time. In most cases, those were the only parts of their bodies hard enough to survive decay and fossilization. But bone, with its hardened mineral component, often persists where other tissues fade away. If bone had never evolved—or if our ancestors had skeletons made of different materials—the only records we'd be left with would be those of rare, exceptional preservation. Step back to simply admire the beauty of bones—the organic architecture of a body's internal structure, each element having a character all its own—and it's no wonder that we display the ancient dead in museum halls like works of art. Da Vinci couldn't do better.

It would be convenient if the story of how our skeletons came to be was a linear progression of accumulating parts. But that's not what happened. Even though animals like *Pikaia* set up the basic format of the vertebrate body plan, the origin of bone—and its expression of the tissue into a supportive skeleton—is a circuitous story that develops from the outside in rather than the inside out. The tale makes sense only through the lens of evolutionary happenstance, with some of the earliest fish swimming into focus to introduce us to some of the world's first bony skeletons.

The primordial tissue that began to approach what we know as bone evolved around 455 million years ago. This was aspidin. It needs a different name because there's a fundamental difference between this tough substance and true bone: aspidin is acellular. I had a difficult time wrapping my head around this fact when I first learned it. In our bodies, cells are organized into tissues, and tissues into systems. The cell is the common denominator, even in bone. But this is a modern bias. Aspidin was the stuff that bone eventually evolved from, and the material held more structural resemblance to the teeth firmly socketed in our mouths. Aspidin, in the early fish that bore it, accreted like cement. It didn't change or grow like our bones do. Instead it built up on itself as a wall between ancient vertebrates and the outside world.

The ocean was a harsh place when aspidin evolved. There were predators with crushing mouthparts, raptorial appendages, pinching arms, and compound eyes to detect prey flitting through the water column. Armored animals stood a better chance of surviving and leaving their gift of anatomical happenstance to their descendants. Vertebrates missed the chance to evolve flexible keratinous shells when their early precursors split from their last common ancestor with arthropods more than 558 million years ago. Bone was an entirely novel response to the dangers our distant relatives faced, made possible by a glut of $CaCO_3$—or calcium carbonate—washed into the seas by shifting, eroding continents. Life needed armor for protection, and there was enough available raw material to grow it.

Over time, though, bone transitioned from a static tissue into something more flexible, more reactive—something capable of remodeling itself and healing from injury. The first evidence of this can be seen in the group of fish called osteostracans—think of an armored tadpole and you've more or less got the picture of what they looked like—both in their outer skeletons and in the casing around their little brains. And, as it turned out, bone was a hit. The ancient seas soon saw an explosion of strange fish ensconced in bone. Cope himself named some of these, such as the antiarchi. These were armored fish that looked like Roombas with tails sticking out the back, and they sucked up small morsels through jawless mouths. And that's not all. Once there was bone, the path to jaws opened wide. Swimming alongside the antiarchi were the much more formidable arthrodires—fish that could bite and were covered in bony plate armor. The most famous of these is *Dunkleosteus*, a predator the size of a great white shark with shearlike jaws that resembled the world's largest pair of staple removers, but there were also gentler giants such as *Titanichthys*—similar in size to its fearsome relative but with more minimalistic mouth plates that would have relegated it to a life of filter-feeding rather than chomping.

Much like the backstory of bone and the skeleton, the origin of jaws is not a nice and neat tale. Paleontologists and anatomists still disagree about how these essential parts of our own bodies initially came to be—if they are modified from gill arches of early fish or evolved in a different way. But regardless of what the fossil record eventually reveals, a jaw is a pretty damn handy mechanism. The earliest fish just had openings to suck food into. Having a jaw, on the other hand, means that animals can control what enters their mouths. Jaws also assist with breathing, whether that means a shark in repose on the sea bottom pumping its jaws to send water out of its gills or a winded human opening her mouth to take in more air after a run. Marathons wouldn't exist without jaws, and had these hinges never originated, Peter Benchley's novel *Jaws* would have to be renamed something like *Pharyngeal Slit* or simply *Hole*, which doesn't have the same effect.

The basic skeletal elements that could have formed the first jaws were around for a long time, but a few changes had to be made before that evolutionary innovation could pop into existence. The skull had to be reorganized to make room for a movable mouth first, including the separation of the ducts of the nose from each other as well as from a tube connecting to the mouth, called the nasohypophyseal duct. A reorganization of the skull had to come before a bite was even a possibility, but we know a little more about these changes thanks to some recent fossil finds.

The fossil fish *Entelognathus primordialis* is our key player here. Its name means "primordial complete jaw," and *Entelognathus* certainly lived up to it. At 419 million years old, *Entelognathus* lived during the earliest days of jaws, but its archaic anatomy retained some transitional features that illustrate how jaws came to be. The skeleton itself, only about four inches long and preserved in three dimensions, looks something like a fish-shaped turtle shell without the turtle inside it. The fish's entire form is dictated by the shape of its pebbly external bones, and below its blunt snout is a little jaw. Even at such an early date, this fish had a combination of premaxilla, maxilla, and dentary—key bones that make up the upper and lower jaws—seen in bony fish and their descendants, including us. This made *Entelognathus* one of the earliest gnathostomes, or vertebrates with a jaw.

Contrary to the importance we might currently put on our connection to this ancient fish, however, jaws were not evolutionary game changers the instant they evolved. The first jawed fish didn't swish around gobbling up all their competitors. Once again, it was unforeseen circumstance that made the difference. Over time and thanks to a mass extinction that wiped out many of the jawless fish, gnathostomes became the most diverse group of vertebrates around. And once there were jaws, there could be teeth.

The origins of teeth and jaws are rooted together. While plates of bone protected our early vertebrate ancestors from the dangers of the outside world, the archaic versions of the tissues that make up our teeth were there, too. Shields of bone in early jawless fish were often dotted with the precursors of dentin and enamel, which are much harder, more mineralized tissues. (This is what makes your teeth so good for biting.) As jaws formed, teeth came along for the ride. A serendipitous fossil helps show how this happened.

Compagopiscis lived 420 million years ago, right at the great profusion of jaws. This fish was one of the placoderms—the scary swimmers like *Dunkleosteus* that had slicing edges of their jaws made of bone armor rather than traditionally toothy tissue like enamel. But when paleontologists used high-powered X-ray tomography to have a look inside the jaws of a juvenile specimen, they found true teeth, after all. The little spikes had a hard outer coating of dentin and even a pulp cavity to nourish the teeth. Given that the adults lacked these, the conclusion was clear: *Compagopiscis* had teeth all along but they were worn away as they aged, their toothy spikes giving way to a jaw where tough bone took on the brunt of the punishment. The discovery added credence to the idea that, like bone, teeth have their origins on the outside. Earlier bone boxes that qualify as archaic fish had tissues similar to dentin and enamel—the hard tissues that make up the core and outside of our teeth, respectively—poking up from their armor. These spikes didn't move but were a way to bite back against any predators that might try to nab them. As vertebrates began to lose their toothlike armor, these struts remained around the nascent jaw, becoming embedded in the hinge. It took millions of years for them to become permanent fixtures for the whole of the animal's life, but, as baby *Compagopiscis* have shown, teeth soon followed Earth's earliest bites.

By this point, vertebrate life was looking quite a bit like us. You might not see much direct resemblance between yourself and the bones of a carnivorous fish that sliced its meals with jaws that look like a giant set of shears, but many of the basic parts were there: a distinct head, a spine, jaws, and teeth. On top of that, some lineages of fish began to ossify their internal skeletons. This is the long-term version of what goes on in our own bodies when cartilage—and sometimes other tissues—gets flooded with calcium and turns into the actual substance bone. As the first internal skeletons formed inside ancient fish, that base soft-tissue structure set down by protovertebrates such as *Pikaia* began to become hardened along the spine and skull, providing internal protection and support. This is the contradiction held in our bones. The biological building material bone evolved for outer protection before being co-opted for internal scaffolding. Bones evolved from the outside in. Our skeletons are sunken body armor. And there was another change that happened during this critical window, one essential for what I'm typing out right now. Fish evolved fins.

Our complement of two arms and two legs was not predestined. We're so used to this state of being that a quick and easy way to make the humanoids of science fiction and fantasy seem alien or monstrous is to stick on an extra pair of arms or legs. Yet in a plausible alternative evolutionary timeline, I could be sitting here talking about only two or even six limbs as the usual count. There's nothing particularly special about two sets of paired appendages. It just happens to be the number that we have. There's a historic reason we have two arms and two legs; once again, that takes us back into ancient seas.

As different as they look from each other, our arms and legs are extraordinarily similar. Our hands and feet are organized according to the same pattern: flexible digits are followed by longer bones with a mess of smaller bones leading up to the articulation at the hinge with our limbs. In both arms and legs, two long bones (the radius and ulna in the arm and the tibia and fibula in the leg) meet up with a single, sturdy limb bone (the humerus in the arm and the femur in the leg), which in turn connects the appendages to the rest of the body. This similarity goes way back into our ancestry.

The origin of the very first fins is still mysterious, but the fossil record is clear that pectoral fins—the ones flanking the equivalent of a fish's chest—came before those at the hip. Somewhere along the line there was at least one population of fish in which the genetic cues for forming those pectoral fins didn't shut off when they were supposed to. This happenstance of gene regulation formed a duplicate set of fins at the hips. The fossil record reflects this evolutionary leap. In a little fossil fish named *Guiyu oneiros*, found in the 419-million-year-old rock of Yunnan, China, for example, the fins that jut out from the area of our scaly predecessor's pelvis are remarkably similar to those that stick out sideways from its chest. This isn't just superficial similarity. It's a sign of a chance event that ended up opening a vast array of possibilities for life on Earth.

By about four hundred million years ago the standard elements of the vertebrate body plan were already set. We don't just have an inner fish, as paleontologist Neil Shubin has reminded us; we are *still* fish. In the panoramic view of life, there's barely a dime's worth of difference between our skeletons and that of the coelacanth, the deep-sea fish that was thought to be extinct until a freshly caught specimen was dredged up off South Africa in 1938. Differences that might seem significant to us are really only smaller-scale refinements. A skeleton that, until recently, rested across town from what remains of E. D. Cope helps make the point.

My appointment to see the 375-million-year-old rock star was set for a February morning in 2015, not long after the recent Snowpocalypse had locked Philadelphia in a frozen state. I half-walked, half-skidded my way over to the Academy of Natural Sciences, nodding to the poised sculptures of the sickle-clawed *Deinonychus* out front before gratefully entering the warmth of the building and calling up for Academy paleontologist Ted Daeschler. I felt a little jittery, not so much from the cold but from who I was there to meet. Should I ask for an introduction right away? Should I make some small talk and catch up with Ted first? Would I be able to take photos? I wandered around the dinosaurs next to the entrance for a few minutes before I was fetched back to collections, and after some happy chitchat Daeschler asked, "Do you want to meet the poster child?"

The Academy's collections aren't exactly spacious. There's just enough room for the closely packed cabinets to swing their doors open, but not for you to stand in front of them, unless you can make yourself as flat as the technical papers describing the creatures carefully nestled in the drawers. There's a lot of history kept there and it needs every inch. I shuffled to the side as Daeschler unlocked the proper doors, doing the back-and-forth dance sometimes required in collections spaces as he carefully slid out a tray from the middle of the rows. There, resting with a self-satisfied smirk on its ancient face, was *Tiktaalik roseae*.

I had seen casts and restorations of the famous fishapod before. Museums all around the country have their own facsimiles, often posed next to a life restoration flexing its forefins to push its eyes and nose just above the surface of the water like a friendly crocodile. But there's no substitute for seeing the real thing. These were bones that had to grow and change and move as this creature paddled itself about between two worlds, right at the cusp of something unprecedented. It didn't have arms like ours, but the basic anatomical correlates of our arms, legs, shoulders, and hips were already there—what we can recognize as transitional features between the fish who came before and the terrestrial vertebrates who would eventually go on permanent shore leave. Meeting *Tiktaalik* was a joy not because of its status or because of any family relation—there's no way to know whether I should have greeted the fish as my great-great-great-great-and-many-more ancestor or gone with a more informal "Hey, cousin"—but because of the world those bones represent. The ancient environment that molded and interacted with those osteological remnants is lost to us now, but we can still touch it through this creature who unintentionally and unknowingly gave its body to the fossil record.

Enthusing over the fossils, I remarked to Daeschler that *Tiktaalik* fundamentally changed what we thought about one of the most pivotal events in evolutionary history, the time when vertebrates crawled out of the sea onto land. Daeschler could have tooted his own horn here for all the work he'd done in the freezing cold of Ellesmere Island digging up these specimens and the hours back at the museum parsing their secrets, but instead he took a beat and said, "Well, as much as any of the other fossils from that transition have." And, slightly embarrassed by my fanboy moment, I knew he was right. The skeleton of *Tiktaalik* was simply the emblem for a full spread of fossils that had changed the narrative of one of the most celebrated moments in our fossiliferous backstory.

The emergence of vertebrates onto land required some major ecological changes. The land had to be prepared. The plants came first, claiming their spots on ancient shorelines about 475 million years ago. Their story is an epic in itself, the way they evolved from species tied to water into the founders of primordial forests required dramatic changes. Invertebrates were the next to poke their heads above the water line. About 428 million years ago—after plants started to cover the ground, around the time fish developed the ability to bite—small, millipede-like arthropods started to crawl around on the prehistoric shore, and by 391 million years ago there were early insects munching on the prehistoric salad bar. So by the time fish like *Tiktaalik* were poised even to possibly start prowling along the tide line, plants had already been there for 100 million years and arthropods for over 50 million. But vertebrates needed more than just the right anatomical equipment to move to land. They needed a reason, and it turned out to be the same one that drew the invertebrates out of the seas so many millions of years before: dinner.

Fish capable of spending more time on land not only removed themselves from the danger of being preyed upon by their neighbors; they were also able to feast in a world free of competition. There was a smorgasbord of invertebrate food with no one around to eat it. But *Tiktaalik* and similarly equipped fishapods probably didn't spend much time on land at all, if they ever did, while the salamander-like *Ichthyostega* and *Acanthostega* could get around on shore but were much more comfortable in the water. It's only in the span of about 360 to 345 million years ago that amphibious and terrestrial life really took off for the vertebrates. Paleontologists previously called this period Romer's Gap, after the famed fossil expert Alfred Sherwood Romer pointed out the paucity of fossils from this critical time. Recent discoveries have filled in this mysterious span and shown a fantastic radiation of tetrapods in and out of the water.

As Ted Daeschler pointed out to me during my visit to see *Tiktaalik*, any true invasion of the land didn't coincide with the origin of fingers and feet. The time when fish became amphibians and made themselves at home on land happened millions of years later, during a chapter in Earth's history that we're only just starting to get to know. All of this makes me look at my hands a little bit differently. Their ancestral form was not the saving grace of intrepid scaly fish fighting for each breath beneath the relentless sun. They evolved in the water to help my fishy ancestors push themselves around through the muck and pop up to the surface for the occasional exoskeletal morsel. And they opened a world of possibilities for the creatures that followed.

While fish continued to swarm the seas and amphibians evolved to occupy the space between land and water, one group of vertebrates took to living on land full-time. These were the amniotes—so named for the self-contained amniotic egg that allowed them to thrive away from moist shores—and over millions of years, one superficially lizard-like group set the foundation for our own branch of the family tree. These were the synapsids—or protomammals, if you like—and while the modifications they underwent were more subtle than some of the changes we've seen so far, they still have significant consequences for our bodies today. The skeletons of our ancestors were just modified versions of the classic fish chassis, but it's worth looking at some of the additions and subtractions that happened between the time our ancestors went from looking like lizards and when they became fuzzy, weasel-like protomammals. For example, you don't have bones in your eyes.

The heyday of the protomammals was the Permian, between 298 and 252 million years ago. This included the time of *Dimetrodon*—a sail-backed animal often mistaken for a dinosaur but more closely related to you and me—and a bevy of other fuzzy oddities. Many of them, following their reptilian ancestors, had sclerotic rings—circles of small, interlocking bones—suspended in their eyes. *Dimetrodon* had them. The tusked, piglike dicynodonts had them. The saber-toothed gorgonopsians had them. Even the lineage that would eventually spin off mammals—the cynodonts—had representatives with sclerotic rings. But at the origin of the cynodont lineage we belong to—technically called probainognathians, if you want to try that mouthful of a term—those eye bones were lost. Size might explain why.

Most of these weasel-ish cynodonts were relatively small compared to other protomammals. Their skulls were about four inches long, with bodies comparable in size to a modern fox. At such a small size, eyes don't need as much structural support. Big eyes have more trouble with a phenomenon called accommodation, which involves muscles allowing eyes to take in various depths of field to keep the view sharp. In fishlike marine reptiles called ichthyosaurs, for example, deep-diving species that thrummed their tails through the dark in search of squid had huge eyes with equally impressive sclerotic rings to help them keep their prey in focus in the dim waters below. In little animals with smaller eyes, though, the same constraints didn't come into play. There would have been even less need for eye bones if these little insect-eating cynodonts were mostly active at night or in low-light conditions, when taking in a whole wide landscape was unimportant. Not that such changes would have dictated the loss. There are small reptiles and birds that still have sclerotic rings. But the miniaturization of mammals and their ancestors may have opened up the possibility for the eyes of our forebears to go soft. And given that bones can break, this is probably just as well. I'm glad I'll never be making a trip to the emergency room with a broken eye bone.

Ocular osteology wasn't the only loss during this time in our ancestry. As far as the fossil record can tell us, our protomammal ancestors lost their belly ribs and reduced the number of ribs jutting from their spines before the origin of the first true mammals. The center of our chassis is that of a fuzzy, snuffling cynodont. Breathing might be the critical factor here. Remember *Dimetrodon* and the more lizard-like protomammals? They probably didn't breathe with a diaphragm and instead pumped their lungs by actively moving their ribs like lizards do. They also moved more like lizards, their bodies shifting side to side. This would have constrained the ability of *Dimetrodon* and similar protomammals to breathe while running, making a trade-off between running off to eat your meals in peace and actually consuming them. But as our cynodont ancestors and related protomammals moved away from a side-to-side shuffle, carrying their bodies higher off the ground and moving them in more of the up-and-down undulation we still see in running mammals, new evolutionary avenues opened up. They didn't need belly ribs to protect them from the rough ground, nor superflexible ribs to pulse their insides. Their rib cages could be a solid support while the tissues inside did all the moving for them. Our cynodont ancestors were protomammals that were born to run.

All that moving and shaking also caused the earliest beasts to gain one of the strangest bones in our skeletons. I'm talking about the kneecap, the rounded knob of bone right at the junction of your upper and lower leg bones. What makes it so peculiar is the fact that kneecaps are what are called sesamoid bones, or "seed bones" that grow inside tendons. There's no solid attachment between your patellae and the rest of your skeleton. They're held in place between your quadriceps and patellar tendons, right over where your femur and tibia meet. They started off as cartilage when you were a baby but fully changed to bone by the time you turned three years old.

We're not the only creatures with kneecaps. Patellae have evolved at least once in birds, many times in lizards, and multiple times in mammals, almost all of these lineages acquiring them sometime in the Jurassic. That's because a patella is a pretty handy bone to have. Whether in a bird, a cat, or us, this odd bone can act like a lever to support an animal's body weight as it stands and moves. The bone can also help keep legs straight—kneecaps are part of a stay apparatus in horses and other animals that helps them keep their knees locked in a flexed position while still. Yet these are relatively specialized features of modern animals. We know almost nothing about what the kneecap did when it first evolved, when both the tendons that held the bone and the bone itself were thinner and not as well developed. It may have been a matter of crossing a certain threshold when individuals who just happened to have sesamoid bones at their knees were better able to cope with the stresses of running and, as luck would have it, left more offspring to carry on the trend. No one knows just yet, but if the rest of the skeleton is any guide, kneecaps could be another fortuitous happenstance that helped piece together who we are inside.

Protomammals made another contribution to your skeleton that's harder to detect, at least with your eyes. Follow the back of your jaw to that pressure point just below your ears. There, where your ear canal dives into the skull, is an opening called the external auditory meatus. This is the portal to your inner ear, and within that hollow are a trio of bones called the incus, malleus, and stapes.

The three are held deep inside, set up in a kind of Rube Goldberg arrangement. At the end of the auditory tunnel is the tympanic membrane. Think of it like a cupped drumskin stretched over the opening. The malleus connects to the membrane, with the incus following and the stapes after that, all set up in a kind of level arrangement that transmits vibrations—which we hear as sound unless something is wrong—into our ears. This is always going on. The reason I can hear my fingers pushing the keys as I type, Led Zeppelin's "Over the Hills and Far Away" quietly playing from the speaker next to me, and my dog Jet snoring from the nearby couch is these three tireless little bones.

Thanks to our protomammal relatives, we know that our ear bones are modified bits of jaw. Look at the lower jaws of early protomammals—like our friend *Dimetrodon*—and you'll see a puzzle of bones. Following the trail leading to our cynodont ancestors, though, the bones at the rear of the lower jaw get smaller and smaller until the entire jaw is made up of just one bone, called the dentary. What happened to those other little bones, then? Their proximity to the ear made them ideally placed to help pick up the vibrations we interpret as sound, and they modified to fit that purpose at the same time. Jaw and ear evolved in tandem, a wonderful example of how evolutionary modification creates transcendent change. The way our ancestors ate helped provide the catalyst for giving us mammals sensitive ears to carefully listen in on the world around us, not to mention to compose the music that surrounds our lives. Without our protomammal ancestors, there'd be no electric guitar, and who would want to live in a world like that?

But we need to keep moving along our skeletal timeline. The next chapter of life's history—the Mesozoic—saw a mammalian explosion. Sure, the small beasts lived under the feet of fearsome dinosaurs and generally stayed small, but early mammals were no evolutionary slouches. They diversified and proliferated into an impressive number of niches and body types, from ancient equivalents of squirrels and hedgehogs to badger-like species that ate baby dinosaurs. But that's a tale for another time. Where we pick up the story is right at the pinnacle of dinosaur dominance, practically the Tuesday before the asteroid struck what's now the Yucatan Peninsula, when the first primates appeared.

In truth, no single animal can claim the title of prime primate. Instead there were populations of mammals whose genes and anatomy were constantly in flux, individual varying from individual and population from population. Still, it's possible to follow the trail of petrified crumbs back to when primates diverged from all other mammals in the form of a creature called *Purgatorius unio*—a fitting name for a mammal that suffered through the end of the dinosaur's reign, earning the heavenly reward of a wide new world where its kind would climb and scamper through the humid forests.

Precisely what *Purgatorius* looked like isn't clear. It's only known from fragments first found on Purgatory Hill in Montana. Yet, while meager in terms of skeletal completeness, the teeth, ankle bones, and other fragments reflect some important things about *Purgatorius* and its relevance to how our bodies are laid out.

The ankles of the ur-primate were already suited to a life skittering and jumping between branches. This dovetailed with the evolutionary expansion of trees that bore tantalizing seeds and fruit. Dinosaurs aside, this was paradise for primates. And while most of the *Purgatorius* skeleton still awaits discovery, its place between insect-eating mammals and the somewhat squirrelish species that followed it—technically known as plesiadapiformes—has typically put paleontologists in mind of something not unlike a tree shrew. Imagine *Purgatorius* scampering along a branch heavy with fleshy, ripened fruit, clenching its hind feet as it reaches down with its dexterous little paws to brace itself on the sweet-smelling flesh before nipping off a snack to carry to a less conspicuous location.

If you could see this protoprimate in life you probably would write it off as a long-snouted squirrel. Primates had diverged from other mammals, sure enough, but the key traits that speak to us on a visceral level—grasping hands with the essential opposable thumb and forward-facing eyes that gaze at you as you gaze at them—had not yet evolved. *Purgatorius* and primates like it only set the base conditions for these traits to emerge from the various anatomical possibilities, and it was life in the trees that gave our ancestors these items of inheritance to pass down to us.

All modern primates have forward-facing eyes. This is one of the marks of our family, and a feature we share with many of the predatory animals that stalked our ancestors. That's because forward-facing eyes grant us binocular vision and the ability to better gauge distance—sort of a useful trait to have when you live in the trees. Every habitat is three-dimensional, but some are more so than others. A life leaping and clambering through branches requires an adept ability to judge distances, lest a primate fall to injury or death. More than that, early primates were probably insect snatchers, and the way metabolism works would have required that these small, energetic mammals frequently catch food. The smaller you are, the faster you burn through calories, and as small primates today such as tarsiers and mouse lemurs show, insects are an excellent source of high-energy food. This doubled down the pressure for binocular vision. Needing to nab fast-moving prey in a tangled environment provided the impetus for primates skilled at telling exactly how far the next branch or a tasty beetle was, and it also might explain the hands that would unlock the world for us. If early primates were anything like their living counterparts, they didn't leap after prey mouth-first. They would have stabilized themselves on branches with their feet and tails, grabbing at flies and beetles with their hands before gnashing through their victims' exoskeletons. Articulate hands would have let them pluck fruits and leaves, as well as given them a better grip. While only distantly related to us as far as primates go, a mouse lemur trying to snatch moths out of the air is an echo of what our Eocene ancestors were doing.

Our eyes that tell us so much about ourselves and each other, and the hands we've used to literally shape the world for good and ill, were already in place by fifty-five million years ago. All you have to do is take a stroll through any reputable zoo's primate enclosures to see this. From the ring-tailed lemurs to the lowland gorillas, all of them share these traits in common with us. These traits were so obvious, in fact, that the founder of modern taxonomy, Carolus Linnaeus, put us all together in the family Primates a century before Darwin's and Wallace's evolutionary theories would make sense of those similarities. Pick different traits and you can go further and further back in time. Our ear bones and lack of belly ribs will take us back to the cynodonts. Our fingers and toes are owed to pioneering landlubbers. Our jaws are thanks to twitchy fish. There is no single moment when our bodies became distinctly human. *Homo sapiens* is a process, not an end point. We will no doubt continue to change, and to see why we need to abandon the grand view for a moment and zoom back down to the smaller, frenetic world inside our bodies. A tar-stained skeleton will help us focus.

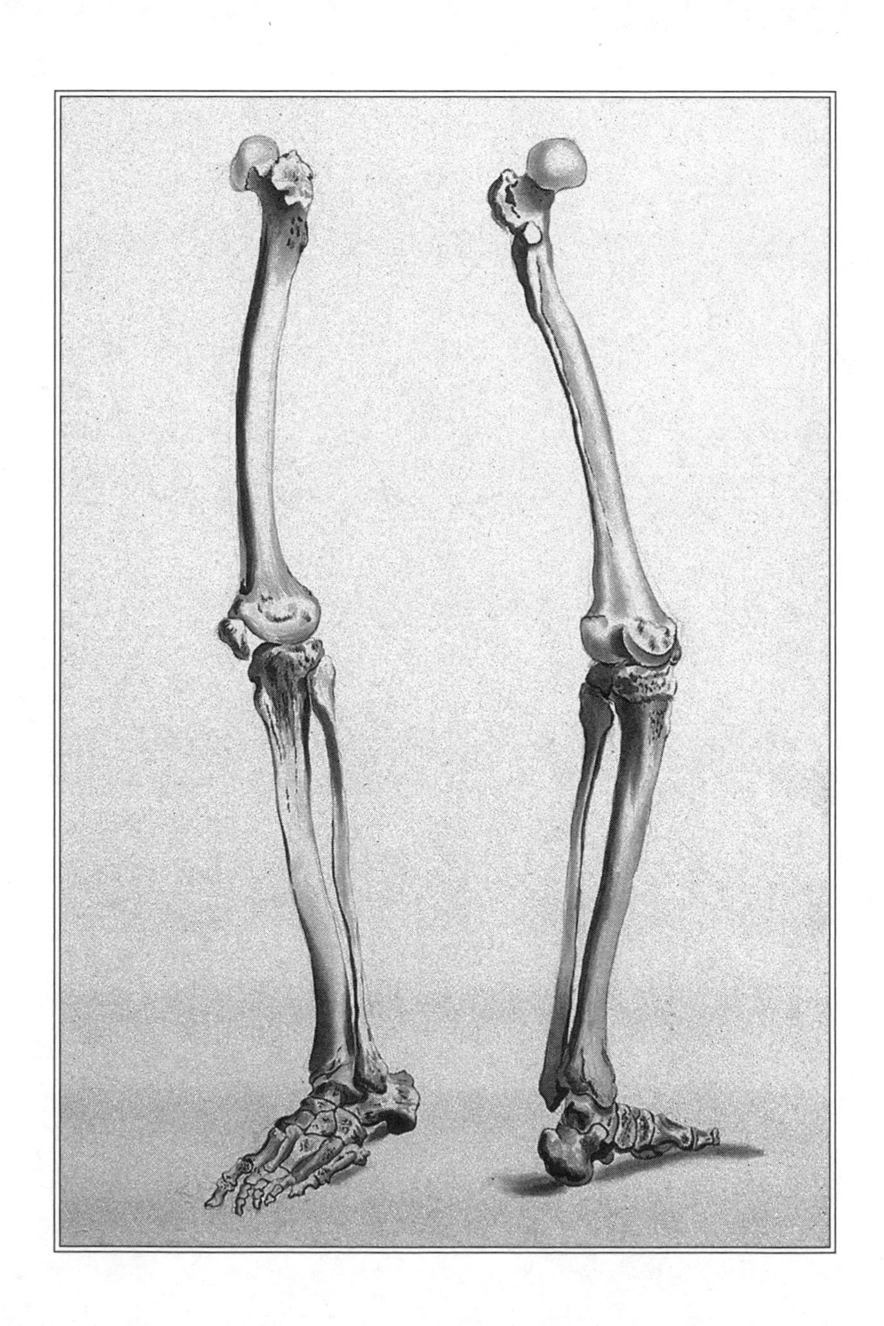

THREE

A BAG OF BONES

You smell La Brea before you ever see it. The scent of asphalt clings to the air around the bustling L.A. block on your approach, and if you're content to roll down Wilshire Boulevard without looking at the black bubbling lake next door to the modern art museum, you could almost be forgiven for thinking the acrid vapor is emanating from a road crew trying to patch up what traffic and California's frequent tremors have cracked apart. I say *almost* because I can think of no reasonable excuse to pass by the richest fossil graveyard on the planet. The trio of mammoths along the shore of the flooded excavation pit out front, a long-tusked adult impassively standing by as a youngster vainly reaches its trunk out to its floundering mother, will let you know that you're in the right place. Turn left at the screaming elephant. You can't miss it.

La Brea's asphalt seeps were not initially discovered by paleontologists. (I say La Brea's *asphalt seeps* because, as friend and museum volunteer Herb Schiff once pointed out to me, the title "The La Brea Tar Pits" translates to "The the Tar Tar Pits." This hasn't stopped marketers from recently renaming the site the La Brea Tar Pits and Museum.) The first record of the site goes back to explorer Father Juan Crespi, who in the August heat of 1769 was impressed by the "extensive swamps of bitumen"—or asphalt—in what the Spanish officially established as Los Angeles twelve years later. This was a huge pool of tar for construction projects that would follow, but there was also a danger to this aromatic place. According to José Longinos Martínez, "In hot weather animals have been seen to sink in [the asphalt] and when they tried to escape they were unable to do so, because their feet were stuck, and the lake swallowed them." It's hard to think of a worse way to go than suffocating in this inky trap.

Sometimes, Martínez wrote, bones were worked back up to the surface by the methane that burbled and farted from the site. But the bones were altered. The exhumed osteological tidbits looked petrified and were stained a rich chocolate brown. No one really understood the truth of those bones, and no one seemed particularly interested in finding out anything more for about a century afterward. Common sense said that the bones offered up by the mire were clearly the disassembled remains of wayward wild animals and livestock that careless farmers had allowed to wander into the pits.

It wasn't until 1875, when geologist William Denton visited Rancho La Brea, that hints of the site's deep history began to seep to the surface. The chief clue was a broken, curved fang presented to Denton by mine operator Major Henry Hancock (for whom the park surrounding the site is now named). This tooth, which Denton estimated must have been eleven inches long when complete, could not have belonged to any living animal. The weapon must have come from a cat that prowled the earth during the last Ice Age and had only been formally recognized by paleontologists in 1842. The tooth was one of the killer canines belonging to *Smilodon fatalis*—the last of the great saber-toothed cats, and the mascot for the museum that now collects, cleans, and researches every bit of Pleistocene life that is drawn out of the tar-soaked sediment. The staff's attention to detail is so fine-tuned that they could throw open all the drawers tomorrow and start reassembling the bone bed almost as it was in the ground if they wanted to, right down to the last twig and beetle carapace. With that first fragment of canine, Denton had started to peel back the history of what is the richest and most important fossil site on the entire planet. Shallow puddles of black goop that continue to ooze out around the park grounds are a small glimpse of what has been going on for hundreds of thousands of years.

That ancient history includes humans, too. People knew of La Brea long before the paleontologists, the miners, and the explorers. Even though the site became famous for its unrivaled slurry of dire wolf, saber cat, mastodon, and other megafaunal bones, traces of humanity have been exhumed from the same pits. Some of them are from relatively recent times. Just a few thousand years ago, after the extinction of the mammoths and saber cats but long before the Spaniards claimed the region as their own, the Chumash people of southern California used the available asphalt to seal their bowls, bind fishing lines to hooks, and otherwise stick together or seal what they made. Their tools and products of industry are sometimes found in the upper layers of the pits. And there's an even older sign that this strange place held special significance for some of

America's first inhabitants: the only human skeleton to ever be pulled from the tar.

The body was an early discovery. In 1914, while working on a pocket of the site called Pit 10, paleontologists uncovered a small aggregation of human bones. None had been found there before, nor have any since. Not only was this unprecedented, but the researchers could confidently conclude that the tar-stained bones—including the skull, a few vertebrae, ribs, hips, and a femur—belonged to a single individual. With the exception of some recent finds made when the neighboring Los Angeles County Museum of Art expanded their parking space, such as Zed the mammoth and Fluffy the American lion, most bones found at La Brea are isolated elements that have stuck and slid in the pits for millennia. In the tens of thousands of years since their Ice Age owners initially became interred, the bones have disarticulated and disassociated, eventually becoming jumbled up into what looks like the world's messiest version of pick-up sticks. For human remains to not only be found, but found together, was a shock.

You won't find the remains on display. At least, not the real ones. There's a replica of the femur—the original of which was totally consumed to come up with a radiocarbon time frame for the body in the days when geologic dating was in its infancy—tucked away in a side cabinet of the museum that most visitors neglect to investigate, along with a human skeleton cast on the wall of the nearby Natural History Museum of Los Angeles County that was modified to fit the same stature and appearance of the La Brea skeleton. (Look for the places in the long bones that have been cut and adhered to modify a cast into the proper proportions.) Instead, what's left of this person is carefully nestled into a slab of museum-grade packing material safeguarded in the scientific collections, each bone cut into a little cove to keep it from jostling around. Former Page Museum curator John Harris was kind enough to bring the remnants out for me during a behind-the-scenes visit to the institution. Each piece of the skeleton carried the same beautiful La Brea brown color as the saber cats and sloths that made this place famous, and the fact that the jaw was positioned just in front of the skull gave the impression that this tar-steeped person was about to start speaking ancient secrets at any moment. If only it could. For one thing, it'd be helpful to know what happened to the rest of the skeleton.

Let's consider how many bones we would expect to be present in an adult human. From a starting point of about 270 individual bones at birth, the components of our bodies gradually fuse until we wind up with an adult complement of about 206. This sum is the same across our species. There's no difference at all between sexes, a fact that Sir Thomas Browne pointed out way back in 1642 when he noted that the biblical gift of Adam's rib to Eve did not actually translate to a difference in rib count between men and women. But the total number of bones in a skeleton does vary from person to person. Some of that statistical squishiness is attributable to Wormian bones. These are extra scoops of bone that form around some of the sutures in the skull. So if you have a Wormian bone around the lambdoid suture—the squiggle that runs along the back of the skull between the parietal and occipital bones of your cranium—then you're at least one up on everyone else. You may also be a Peruvian mummy; this feature is sometimes called the Inca bone for being so commonly seen on skulls from that extinct culture. Then again, maybe you lack some of the sesamoid bones. These bones, like your kneecaps, sit in tendons, but some of the smaller sesamoids aren't present in everyone. You might be missing the pair of little nubs found in the tendon that runs along your index finger or those embedded in the tendon along your big toe.

What's standard for a skeleton, therefore, is more of an average rather than a set number, and of this complement we have relatively few from the La Brea skeleton. There are the twenty-two bones of the skull, including the lower jaw, but only twelve from the rest of the body. Most of this person's ribs, the majority of their vertebrae, and all the fiddly little bones of the hands and feet that significantly up the skeletal count were never recovered.

What happened to all the other parts is a mystery lost to time, one of a number of puzzles surrounding the fossils, but a few things are clear. Every skeletal fragment holds records of a life lived. This collection of pieces contains enough clues that it is among the few ancient skeletons to have a popular audience name. These bones—carefully stored in the museum collections and known to specialists as LACM HC 1323—are known as La Brea Woman, and they are the remains that are going to introduce us to some of the biological basics of our bones, from the broad scale to the minute.

But let's talk about that name, La Brea Woman. Naming is a complicated matter, and one that can hold great power depending on who's choosing the titles. While LACM HC 1323 is a perfectly acceptable label for scientific papers, it also places the human bones found at La Brea in context as scientific objects of study rather than as a person. This is where popular titles often come in. Various human skeletons found throughout the depths of time have been given labels like Nariokotome Boy and Kennewick Man, with La Brea Woman being another on the skeletal celebrity list. So long as osteologists have the right bones—which we'll get to in a second—it's relatively easy to sex a skeleton, and this is often expressed as an accurate reflection of their gender. But in the process of learning and writing about these remains, I've become less and less comfortable using the popular terms. Identifying the sex of a skeleton is one thing. Attributing an entire complex of appearances and behaviors based on assumed gender is another.

Sex, gender, and sexuality are all different concepts with various origins and cultural meanings. They are interlocked but not interchangeable, yet this is often forgotten when we look at old skeletons. Using labels like La Brea Woman automatically creates a range of expectations about a person we actually know very little about, and whom we cannot ask for their interpretation. We can't know how they would have identified their gender or what the nature of their relationships with other people were, making it all too easy for the modern viewer to impress their own values and views onto another person. Often this ends up saying more about the observer and their culture than the bones themselves. In 2017, for example, the remains of two people who appeared to be embracing during the Vesuvius eruption of AD 79 were found to be not female, as previously thought, but chromosomally male. British tabloids immediately ran with headlines calling the people gay lovers, despite the fact that we have no idea about their genders and the nature of their relationship. This is hardly the only time osteological misattribution has occurred, particularly with skeletons found buried together or in a manner that's thought to be contrary to expectations—whether they be warrior princesses or osteological males buried in ways perceived by us as more feminine. One of the most infamous examples is the Red Lady of Paviland—a skeleton found in England that the nineteenth-century naturalist William Buckland identified as a young prostitute based on the presence of shells and ochre, but that later turned out to be a young osteological male. A salacious story overshadowed more fundamental observations, as still happens today. When associated or even entwined skeletons have been discovered, anthropologist Pamela Geller writes, their identification as being "romantically entangled, compulsorily reproductive, or occupationally divided say[s] more about our present state of soci-sexual affairs than . . . about past interactions and intimacies." We must always be aware of our tendency to fit old skeletons into our present systems of ideas and values. We have to be willing to admit how far the evidence goes, and what we can't know about a life lost.

Nor do I feel comfortable claiming bones can reveal someone's biological sex. Our minds and how we perceive ourselves are part of our biology as certainly as our bones are, and to say someone is a man or woman based on skeletal anatomy alone erases what they would say about themselves. What a skeleton *can* show us, then, is osteological sex—a determination of whether a body was physiologically male or female based upon the shape of particular bones. This creates a new challenge. Pronouns matter, and in the absence of positive evidence I'd prefer to use "they" and "them" for people whose gender identities we don't know. But the stories of these people are intertwined with the modern researchers studying them, so relying on "they" and "them" can also cause confusion as to who I'm talking about. To that end I've attempted to eschew gendered popular labels like La Brea Woman—as I feel anthropology and archaeology should do in these cases—but have retained the use of "he" or "she" to talk about the osteological sex of a skeleton whose gender is unknown. And how is this distinction made? There are particular bones that anthropologists can rely on to suss out osteological sex, only they are probably not where you'd expect them to be.

On the surface, you might think that skeletal sex would be simple to sort out. The physiognomy of someone's face, more than anything else, often acts as our guide. Osteologists have made a short list of facial features more associated with females versus males. Males are supposed to have more robust profiles, with squared-off chins and more pronounced ridges of bone over the brow. Anyone cast as Superman or Batman typically has this look, especially in an age when our superheroes sulk and brood more than ever. Females, by comparison, are said to be more gracile in form and lack the suite of features we typically associate with the musclebound. But the truth is that we're an incredibly variable species with almost no sexual dimorphism. We're not like our early human ancestors or modern primates such as gorillas, in which there are obvious, consistent differences between males and females. Many males lack the osteological macho look, and there are some females who have deep chins and other features usually associated with males. There isn't a strict and crisp osteological breakdown of females and males when it comes to skulls. Unless a skull comes with demographic data, assigning it to a man or a woman is, at best, an educated guess.

To more reliably sex a skeleton, then, you have to look elsewhere, and the telltale bones for identifying osteological sex are in the pelvis. At the back of the hip, on the upper flange of bone called the ilium, there's an indentation called the sciatic notch. It's consistently narrower in osteological males and wider in females. And at the floor of the pelvis—where the two pubic bones meet in the front—the two halves meet at a wider angle in females and a narrower one in males. The difference has to do with the capacity for childbirth. Osteological males can get away with narrower hips because no baby has to painfully squeeze through that space. Over the course of time, however, evolution has modified both the skulls of babies and the hips of their mothers to allow our species to keep reproducing itself. And that's why LACM HC 1323 is called La Brea Woman by some. Half of her hip managed to survive burial in Pit 10 and provides the critical evidence that, skeletally speaking, this person was female.

Knowing that LACM HC 1323 is an osteological female, there's always been just enough detail to try and restore what she looked like in life. The Page Museum did that in a potter's box where a bare-chested sculpture shifted back and forth with a reconstruction of her short-statured skeleton. The display was removed a few years back to make room for a fire exit. And in 2009 the museum faced controversy when a volunteer with a talent for forensic artwork tried to re-create what the original owner of La Brea's human bones looked like and published the image online. These artistic efforts are so controversial because they start to get at the contentious identity of America's earliest people. It's the same for a person as it is for dinosaurs or other ancient beings that artists try to restore in ink and paint. Drawing them turns them into something more than bones. They become more real—more like what we know they must have once been—and this can sometimes bring up uncomfortable questions about the ancient dead. In the case of the person entombed at

La Brea, it's the nature of her relationship to today's Native American people.

Exactly who this person was, or what group of people she called a family, remains a mystery. She was undoubtedly a Native American. There was no one else around in California as the last Ice Age ebbed. But, at about ten thousand years old, her bones are too ancient to confidently associate to any particular population or culture. Genetic testing would help, but, given that the process of removing tar from the bones would destroy the genetic material inside, there's currently no prospect of extracting the DNA strands that would help us navigate her kinship. All that experts have to go on is external anatomy. Informative as bones are, this is one area where they lose some of their power. Race is a construction of the living; there's no conclusive way to tie bones to skin color or the broad racial divisions we might think are so apparent looking at the people around us. Our skeletons hold no inherent sign of race, even as we still try to navigate racialized social categories like black, white, and Native American.

We'll return to the problematic analysis of gender, sex, and race in considering the complicated afterlives of bones and drawing out the identities of the dead. But La Brea's skeleton has some other lessons for us, preserved deep down in bones that changed and grew just as ours do. From here, let's zoom in a few levels to better understand those features we all have in common, uniting us as bearers of the remarkable struts, levers, cups, cages, and swivels that make up our osteology.

What exactly is bone tissue? What makes it different from other stiff organic materials, such as the tough chitin of a blue crab's shell?

From a biochemical perspective, bone—in La Brea's skeleton, in yours, and in any other vertebrate's body—is pretty simple. It's a combination of two different materials: a protein part called collagen and a mineral part called hydroxyapatite. They aren't combined in equal measure. Collagen is very common in your body, present in everything from your skin to your tendons to your bones. This is the flexible part of bone, which ensures that your bones have some give and can respond to stress without instantly shattering. It's also pretty stubborn stuff. Paleontologists have been able to tease fragments of ancient collagen out of *Tyrannosaurus rex* bones, for example, meaning that shreds of dinosaur collagen managed to survive over sixty-six million years. *Resilient* doesn't seem like a strong-enough word for a material that can last that long.

Collagen makes up about 90 percent of bone tissue, but it wouldn't really be very useful to us on its own. A classic home science trick can demonstrate why: Leave a chicken bone soaking in vinegar for about three days and it'll be so pliant that you can tie it in a knot. The mineral part of the bone will have been eroded away by the acetic acid, leaving only the wobbly collagen behind. Trying to walk with such a skeleton would quickly have us going sideways, if we could even stand at all. So here's where the other major component of bone comes in. The mineral hydroxyapatite adds strength to the flexibility of collagen, making up about 70 percent of a bone's weight despite being present in comparatively smaller amounts. But we wouldn't want too much hydroxyapatite. Remove the collagen from a bone and it might as well be a crumbly chunk of stone, turning to dust at the slightest insult. Collagen makes bones flexible while hydroxyapatite makes them strong and rigid enough to be biomechanically useful. Take away one or the other and so many fantastic organisms—including us—never would have evolved.

The versatility of bone isn't only in its biochemical makeup. The way bone tissue forms and the way our individual bones grow have also opened up a great deal of biological possibilities. That's because bone is a busy substance. It may seem like pretty static stuff, but it's actually incredibly dynamic. Our bodies grow and change dramatically throughout our lives. These grand transformations rely on the interactions of an entire maintenance crew of specialized cells that grow, maintain, and break down your bones. Among the most important players are osteoblasts. These are the cells responsible for making new bone tissue, coming together in small clusters that lay the groundwork of our skeletons bit by bit. Imagine clumps of cells working like a 3-D printer, laying down layer after layer of bone tissue, and you've more or less got the idea of what they're doing right now, inside your body.

Osteoblasts ooze out a material called osteoid, a kind of pre-bone tissue that's rich in pliant collagen. The osteoid forms tiny crisscross struts around the osteoblast that look like biological latticework. Then, with the osteoid in place, the hard, hydroxyapatite part of the bone starts to precipitate from the biochemical mix of calcium and phosphate rounded up by the osteoblast. The mineral portion of the mix seeps through the lattice, permanently trapping the osteoblast inside a cage of newly formed bone.

And at this point, the bone-forming cells shift gears. The osteoblast changes and becomes comparatively inactive, the skeletal equivalent of painting itself into a corner. All walled up, the osteoblast is now what's called an osteocyte. You've got about forty-two billion of them in your skeleton, and once an osteoblast turns into an osteocyte it hunkers down to play a regulatory role in the slow turnover of bone, having more to do with managing how bone is broken down rather than its buildup. These teardown operations are primarily carried out by a different type of cell called an osteoclast. Close up, the process might look something like what happened to the deck of the *Nostromo* in *Alien* when the face hugger's acid blood melted through the floor. While osteoblasts form new bone, osteoclasts eat away at existing bone tissue. They do so by secreting acid that dissolves the mineral part of bone and an enzyme that breaks down the collagen. This is called bone resorption, and it's an essential part of the regular maintenance your bones undergo. All the major changes we see are because of these little cells, just as mountains rise and fall thanks to the ever-present, gradual forces of uplift and erosion. These cells are the living testaments to the vibrant nature inherent in our skeletons. They formed the bones of La Brea's Pleistocene person, and every other human, just as they did yours.

So now we know something about what bone is made of, and how bone tissue forms. But this isn't osteological anarchy. All these small activities happen according to broader patterns that knit our skeletons together. The way bone forms, and where, is different at varying points in our lives. Parts of our collarbones and some of our skull bones, for example, form through what biologists call intramembranous ossification. While we're developing embryos, the bone tissue that builds up these elements forms within a temporary kind of soft tissue that's a precursor to several of our body's essential systems and fluids like blood. But most other bones in our bodies are formed by what's called endochondral ossification. That means the initial bones of the skeleton are made of cartilage but are eventually replaced by bone tissue as all those little osteoblasts go about their tireless work. The way it happens actually sounds more than a little sci-fi, when you stop to think about it. Blood vessels from our developing bodies poke their way into our cartilaginous bones, making points called nutrient foramina, where the transition to bone starts and spreads. And it gets weirder with the long bones of our limbs. These bones are surrounded by a tough membrane, with the bone-growing cells going about their work in the space between the bone surface and that tough membrane. They lay down bone layer by layer, with bone tissue gradually replacing the cartilage. And as these long bones take shape, those osteoclasts eat up bone on the interior surface even as new bone is being created on the outer surface. The result is an element made of bone tissue with a hollow inside. And much of this happens in our earliest days. Our skeletons are perhaps never busier than when we're still developing inside our mothers. At around eleven weeks before birth, biologists estimate, there were approximately eight hundred distinct centers of ossification in your body where softer tissues were being transformed into bones, the process finally coming to an end when those various centers fused into the roughly 206 distinctive parts of our skeletons in adolescence.

Despite the long time frames bones can survive—thousands of years in the case of the human bones found at La Brea, and longer still for many of the prehistoric species we've met along our journey so far—bone tissue itself isn't very permanent when it's alive. Our bones largely stay in their articulations, yes, but it's not as if your bones ossify during infancy and then never change. Even when you've reached your adult skeletal shape, the osteoblasts keep laying down new bone material just as the Pac-Man-like osteoclasts go about their ravenous work of removing old bone cells. In fact, the balance can shift late in life, leading to conditions such as osteoporosis, in which either the osteoblasts aren't making enough bone, the osteoclasts are consuming too much bone, or both together. All this frenzied and continuous activity always happens on a preexisting bone surface. We're always being added to and subtracted from along the outer surface of our bones, with all that activity taking place between the bone surface and a tough external layer of living tissue called the periosteum. This is the biological wrapper that surrounds your individual bones, with the exception of the patches of joints along your arms and legs. It does lots of different jobs—such as helping blood created by the marrow of our bones get to the rest of the body—but in the story of bone formation, the periosteum generates the cells that become osteoblasts and build up bone on the outer layer.

The result of all this growth and rearranging is a very active tissue that is incredibly versatile, preparing our bodies for all sorts of stress and strain, tension and compression. Part of this strength comes from the way collagen is arranged in bone tissue. It's not just splorted out wherever. Collagen fibers run in different directions in successive layers of bone tissue, going at various angles in relation to each other in the organic latticework. If all those collagen fibers ran in just one direction, bone would be exceptionally strong against one direction of stress but would easily snap if damaged from another angle.

We have three-dimensional lives, with the inner structure of our bones having to deal with various shocks and rattles from all sorts of directions. Think about a femur. (I won't say *your* femur because of what we're about to do in our little thought experiment.) If the collagen fiber of the femur were all arranged in horizontal bands from bottom to top, the bone would easily break if you struck it horizontally—like with the swing of a baseball bat. Arrange the collagen fibers vertically, however, and the bone would be resistant to that horizontal bat strike but extremely weak against a vertical strike, swinging that bat down as if it were an axe. It's the same principle that I learned back when I was taking tae kwon do classes as a kid: to break a board, you want to strike with the grain, not against it. Thankfully for us—and all other vertebrates—the arrangement of our bones has fibers running in different directions throughout, making them much less likely to break. Bones will vary in their makeup of collagen versus hydroxyapatite depending on the forces they endure, but the very way these parts are laid down helps our bones bear our weight and keep us moving.

From here, let's zoom out a little bit from the microstructure of bone to what can be seen with the naked eye. Bone tissue isn't just flexible in the sense of biomechanics. Bone can be organized into different forms throughout the body. The internal structure of the bones in your skull isn't the same as that in your leg bones or your vertebrae. Broadly speaking, our bones usually show one of two types of layouts. There's compact bone—which, like the name suggests, is densely packed with bone cells—and spongy bone that looks more like a network of struts arranged on a more open construction regime. These two different categories of bone offer differing advantages based upon where they are. The bones of our arms and legs, for example, often have more compact bone inside their shafts because these bones need to be strong. They are the elements that dictate how we move around in the world. But the bones in your skull are knitted together and aren't doing much moving around on their own. (At least, let's hope not.) That's why if you look at the cross section of a bone from your skull roof you're likely to find two outer layers of dense compact bones with spongy bone in between. The spongy bone acts as a kind of shock absorber between the harder protective layers, making the bone more difficult to break or shatter if struck. The two types operate in conjunction with each other, depending on the compromises between strength, lightness, and flexibility that are required.

It can be difficult to appreciate all this shuffling that's been going on inside us since our embryonic days. We can look at photos through the years and recognize when we started to get taller, when we began to curse the blight of pimples, when shaving whatever relevant part of our anatomy started to become a question that required answering. But there's no record of the bones beneath. Even when the whole trajectory of our skeletal growth is laid out before us, skeletons of the departed tracking our osteological alterations from birth through old age, the transformation is so familiar as to be mundane. So let's take a moment to consider some of the abnormal ways in which bones can be transformed. These are extreme cases that could be cast as pathological, but they demonstrate how bone is a flexible, reactive tissue that can be molded by what we do in life. These exceptions help demonstrate the biological wonder of bone.

Even though I'm sure many of us would like to forget 2008's *Indiana Jones and the Kingdom of the Crystal Skull*, the movie very briefly touched on a real anthropological phenomenon in the mishmash of pseudoscience needed to get the film's central mystery going. I'm not talking about the crystal skulls themselves. All of the "real" crystal skulls are quartz sculptures created during the nineteenth century to take advantage of widespread interest in ancient cultures and other curiosities. (Recall that this was the heyday of P. T. Barnum and other humbugs, long before the Discovery Channel started trying to again convince people that mermaids are real.) No, at one point our intrepid Dr. Jones holds one of the oddly elongated crystal skulls up to a pre-Columbian painting depicting a person with the same skull shape. And here, fact and fiction part ways. People really did shape their skulls into ornate figures, but it had nothing to do with alien intervention or the need to squeeze a few more bucks out of an aging franchise.

Skulls are not so locked into a particular genetic program that they will always come out exactly the same no matter what we do. Our skulls are composed of various interlocking bones that take years to fully grow, fuse together, and become sturdy. The osteoblasts and osteoclasts that provide the basic biological background of our growth can therefore become co-opted to change the shapes of our skulls given the right times and pressures, the very growth of bone reacting to the squeeze it finds itself under.

As far as human traditions go, intentionally changing the shape of the skull is a pretty ancient one. And popular, too. In addition to the Maya of South America and Native American tribes such as the Choctaw, the Huns and Alans of eastern Europe practiced similar reformation of the skull, as did people in the Bahamas, the Philippines, and Australia, all at different times and for different reasons.

In some places, such as among pre-Columbian people in northern Chile, 88.9 percent of the population have been found to have artificially modified craniums. Some cultures even unintentionally revived the practice. In southern France, in the vicinity of Toulouse, people with elongated, indented foreheads could be seen through the early twentieth century, a sign of a long-standing practice in which a baby's head was tightly wrapped to prevent the infant from fatally banging their head. While the difference in shape had no effect on intelligence, as a more pernicious local tradition held, anthropologists went ahead and dubbed this shape the Toulouse deformity.

What happened to the French babies gets at the basic method of how many other cultures intentionally molded skull shapes into a variety of forms, from rounded to flat to Conehead-type points. These varying skull shapes weren't those of adults who elected to go to extremes in pursuit of body modification, but of people who had these shapes decided for them as children. Our bones are never more active than when we're young, and this is the point at which our skeletons are the most malleable. Part of that's because our skulls take a while to fuse. We are born relatively underdeveloped. While other mammals, like baby antelope, are ready to get up and go quickly after birth, we require years of parental investment and care. As babies, we're all a drag. That's what has allowed us to evolve big brains, though, as our skulls have to be on the flexible side to fit through our mothers' birth canals. The tight squeeze can press on their heads without harming the brain, with the sutures taking years to fuse, knit together, and eventually erase their existence as bone continues to form. The daily machinations of bone, as well as ossification of the temporary cartilage connections, eventually settles the skull into its permanent shape.

Even if they didn't understand the science of this, apparently multiple groups of ancient people figured out that a child's skull is able to be reshaped. So to create these skull shapes adults would bind the skulls of children; the bones, thus constrained, would have to grow into the mold the apparatus determined for them. Some of these instruments looked like bandages or headbands, but others were planks that could be tightened a little bit every day until the skull was the proper shape or the child understandably threw off the apparatus. Each culture had their own devices for achieving the shape that would mark their children as members of their own group.

Why so many different cultures over the past three thousand years or more adopted this practice isn't always known. The simplest explanation is that an intentionally modified skull immediately makes you recognizable as part of a particular social group. That might be a broad community, or it might be restricted to a certain social class. Then again, it might just have been fashionable. As the Huns spread through eastern Europe and Asia more than twenty-seven hundred years ago the people they came in contact with started to modify their skulls, too. This was a fashion wave that flickered across the open steppe, with the various kinds of modification tracking the movements of the Huns across the continent.

Of course, the skull hasn't been the only focus of skeletal modification. The practice of foot binding is another, and I'll refer you to Lisa See's landmark novel *Snow Flower and the Secret Fan* if you wish to pursue those horrific details further. Skulls might be modified with few ill effects, but it's impossible to change something so vital to our existence as our feet, one of the appendages that underwent the greatest change between our ape ancestors and us, without severe consequences. And even though the most dramatic skeletal changes have historically been applied to or inflicted upon the young, we can change our bones as adults, too. In an effort to attain a fetishized hourglass shape, many women across eighteenth- and nineteenth-century Europe wrapped their midsections up in corsets. These weren't the minimalistic and titillating items you'll find at lingerie shops today, but serious pieces of fashion hardware that clamped down hard on the ribs and even harder on the softer parts between the last rib and the hips.

Women of all different ages and classes wore corsets, and the diagrams of how corsets changed the body's internal workings are almost painful to look at. The stomach and liver are crammed down, with the ribs compressed into drooping S-loops. The neural spines of each vertebra, the little projections that stick up from the central body of each bone, are also pushed out of place. Normally they stack nicely one atop the other in a neat midline ridge, but in long-term corset wearers these spindles of bone jut to this side or that, pushed out of their typical positions. Years and years inside the corset significantly modified the skeletons of the women who wore them, creating cringe-worthy shapes. It's no surprise that the corsets of old have been decried as torture devices, cruel manifestations of male-driven beauty standards that ended women's lives too early. But it turns out that tight corsetry was not a ticket to an early grave. Consumption, known to us today as tuberculosis, has nothing at all to do with corseting, despite its continuing cultural association. Circulatory problems attributed to corsets have also been debunked, as have the supposed ill effects of altering the position of the liver. Corsets moved things around to what we might think of as abnormal positions, but they were not death traps. In fact, in a study of the effects of long-term corseting carried out on women with associated demographic data, anthropologist Rebecca Gibson could find no evidence of chronic harm or shortened life span among women with corset-modified ribs and spines. (And while Gibson's study focused on women, it's worth pointing out that there are some osteologically male skeletons that also show signs of corseting and can help track the fashion's rise and decline in various parts of society.) Corseting is another important reminder of just how malleable our skeletons are. Parts of our skeletons and soft tissues are capable of taking on a greater array of shapes than we see even in natural variation.

These outliers underscore the basic processes that are constantly reworking and reshaping our bodies. We are far more malleable than we usually acknowledge. Despite its rigid appearance, bone is one of the most flexible tissues on Earth. It also allows the grace of our every movement. Bone got its start as rigid plate armor on the outside of primordial fish, but as those pieces sunk inside they became an interlocking framework that never shifts by itself and yet, when embedded in flesh, allows for the sensational range of motion our species is capable of.

Now that we know how our skeletons came together over evolutionary time and the processes that formed the bones within our bodies, it's time to turn to how such a remarkable material allowed for the movements both complex and subtle that fill our daily lives. We are, at our skeletal core, slightly modified arboreal apes, just one biological expression of many in a long human history that goes back more than five million years. The transition from life in the trees to a dedicated terrestrial existence is the essential context for why we move the way we do. Lucy, the epitome of a life spent half in the trees and half on the ground, will take us back to the time when our range of motion was being set in bone.

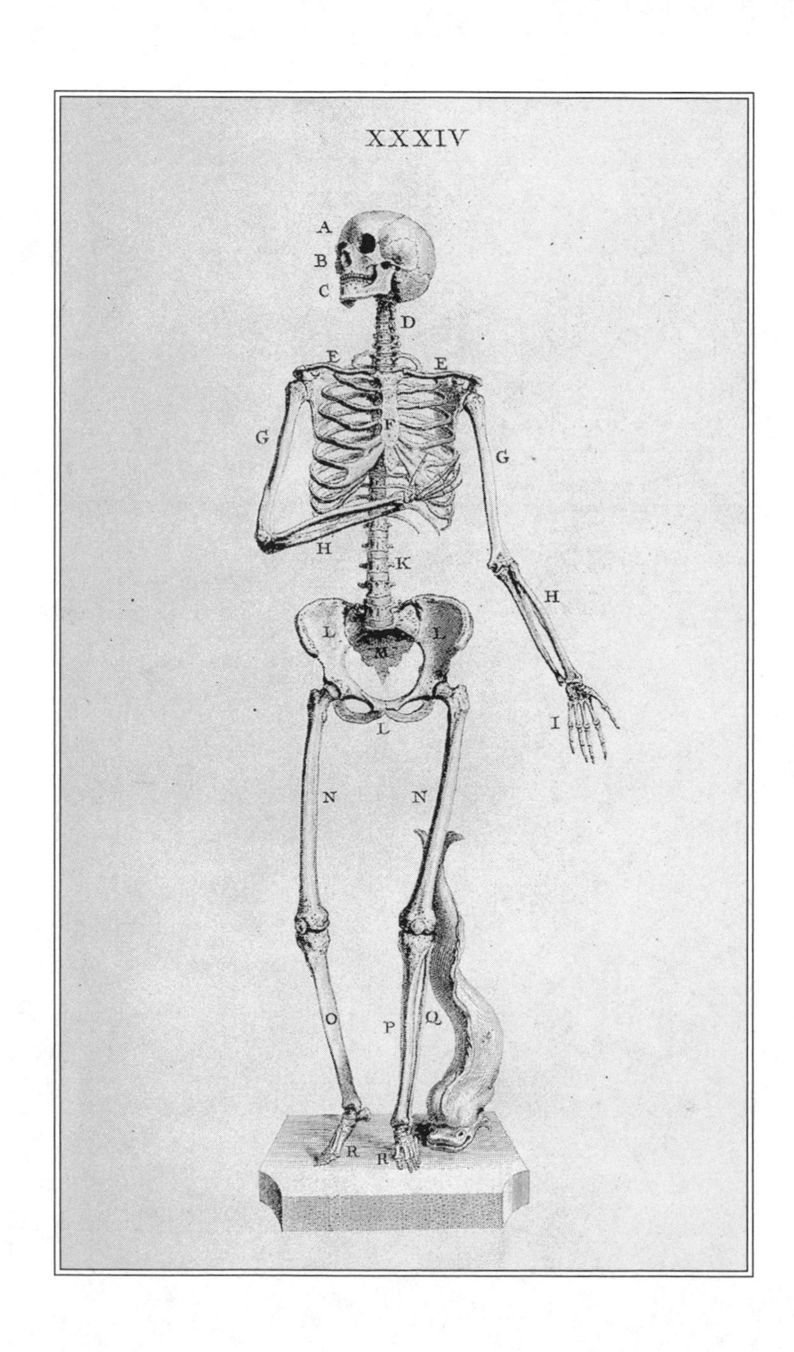
XXXIV
A
B
C
D
E
E
F
G
G
H
H
K
L
L
M
L
I
N
N
O
P
Q
R
R

FOUR

BONE SHAKING

Our skeletons are strange. It might not seem so at first glance. After all, they are *our* skeletons. You and everyone you have ever known are all formed on the same core layout that makes us recognizable as *Homo sapiens*. But that's the subjective view. Look a little more broadly at all the varied shapes skeletons can take and humans stand alone.

In the 3.5-billion-year history of life on Earth, and the six hundred or so million years that animals have been around, anything that we would consider humanoid has only evolved once, and pretty recently, too. There's never been anything like us upright apes. Sure, other creatures have stood up to run around on their hind legs. Dinosaurs and ancient crocodile cousins called pseudosuchians, for example, were the first to do so more than 235 million years ago, back when our own ancestors were still scurrying around on all fours. And while other animals have arms suited to grasping and clambering around in the trees, none possess quite the same osteological arrangements that we do. We're a mishmash of different features, our frames still caught somewhere between our uniquely human trait of having to walk around with our backs straight up and the arboreal ancestry that gifted our species with the dexterous hands that opened up a world of potential. We're not an end point; we're creatures constantly affected and molded by our history. A large part of that history is how movement shaped our skeletons, both in individual lives and over long evolutionary timescales. This is the realm of biomechanics—the way our bodies move. And to get in touch with a critical part of our past, we need to visit with Lucy.

Even if you're not a paleoanthropologist, you've probably at least heard of Lucy. This collection of ancient bones is the most famous fossil human of all time, the titleholder for the name *Australopithecus afarensis*. Other ancient hominins don't share anywhere near the same notoriety. And while no skeleton stands in isolation—their importance can be seen only in the context of one another—the circumstances of Lucy's discovery and unveiling have assured her a permanent place in any consideration of humanity's deep past.

Part of Lucy's fame comes from the historical context of paleoanthropology at the time of her discovery. In 1974, when Lucy's bones were excavated near Hadar, Ethiopia, we knew very little about the critical transformations that changed our more apelike ancestors into being recognizable as people. Lucy sat right at the crux of that change, when confidently striding over the ground was still a new way of getting around, and with about 40 percent of her skeleton known, she immediately became one of the most complete early humans ever found. (That might not sound that impressive percentage-wise, but until then, most early humans were only known from isolated bones and skulls. The old joke was that you could fit the early human fossil record into a shoe box.) And while it's possible for some fossils to be overhyped, it's hard to see how this could be true for Lucy. These bones would prove to be critical for understanding who we are and where we came from. So when Lucy made a visit just across the Hudson River from me, I had to go see her.

I'd seen casts of Lucy before. Any museum exhibit on human origins just doesn't seem complete without one. But while casts are detailed approximations of a fossil, there's something special about seeing the real bones. Replicas, as detailed as they might be, lack the weight of history. It doesn't matter whether we're talking about skeletons, finding the original Famous Ray's pizza, or seeing the Rolling Stones versus imitation Mick Jagger and Keith Richards in a tribute band. A replica just can't match the authentic. To know that the skeleton arrayed before you came from a living, breathing being draws you further back into history. So my choice was absolutely clear. Ticket to the Times Square exhibition in hand, I took the dingy NJ Transit train up toward Manhattan to pay my respects to a person who strode across the planet more than 3.4 million years before me.

Lucy's bones were laid out in the middle of a dim room ringed by casts and illustrations of various hominin relatives that came both before and after her. Each piece, from Lucy's tiny finger bone to the isolated and worn pieces of skull, was nestled gently in a bed of purple foam. Lucy was just one individual, one member of a varied species, like any of us. I wonder how she would have felt about becoming the emblem for an entire extinct species. But while all those yellow bones arrayed in front of me were familiar—humerus, hip, femur, rib, and so on—they came together in a different way than in any living human. With our hindsight, we can see that Lucy was between two worlds—caught between the trees and the ground.

Lucy wasn't a very tall human. The first time I saw one of the restorations of what she looked like in life, at the American Museum of Natural History's Anne and Bernard Spitzer Hall of Human Origins, I was shocked by how a human with such a massive reputation could be so tiny. In that display, eyes scanning the horizon and holding hands with another of her kind, Lucy stands only a meager three feet, seven inches tall. Prior to that point all the images and restorations I had seen had shot Lucy straight from the front or angled up, making her seem larger than life. The bones told the same story. If I lifted off the glass and carefully plucked up each piece—which I was not crazy enough to try to do—I could have easily cradled what remained of Lucy's skull in one hand.

A quick scan of the rest of Lucy's skeleton tells us a few other things about her life, too. She lived at a time before humans had figured out how to use stone tools to enter the carnivore guild, before humans started using specifically manufactured implements to overcome our meager frames and compete with the lions, hyenas, and saber cats for game. Lucy still had more of a vegetarian diet. Her rib cage was not a rounded cask like ours, but more of a funnel—narrowed at the top and expanding out to help house the fermenting vat of a gut Lucy needed to digest the fruit, tubers, and leaves her species relied on as the core of their diet. Her hips flared out wider to help support her stockier build as well, acting as an osteological cup to help hold all those guts in place. Lucy's limbs are different from ours, too, arms proportionally longer with fingers still relatively curved—suited to both manipulating objects as well as clasping tree limbs. But perhaps most remarkable of all is Lucy's spine.

There's no single trait that defines humanity. Our beloved big brains, for example, aren't the largest in absolute or even relative terms. Blue whales have cerebral matter fitting their stature as the largest animals of all time, and our own Neanderthal relatives had brains just as big as ours, as do capuchin monkeys when considered at the same scale. But paleoanthropologists have traditionally looked at walking upright as the hallmark of humanity that separated us from other apes. It's easy to see why. All apes *can* walk upright, but it's not their habitual posture. For example, our nearest relatives in the family tree—chimpanzees and bonobos—usually knuckle-walk on the ground, and their skeletons differ from ours accordingly. Their spines are relatively straight, their big toes jut out to act like foot thumbs, and their forelimbs are suited to both moving through the canopy and ambling about on knuckles while on the ground. In fact, their method of terrestrial locomotion is just another biomechanical response to how an ape can move on the forest floor, representing a different evolutionary path than the one our own forebears shuffled down. It's no better or worse. Just different.

But Lucy wasn't like a chimp. Her spine shows a critical change related to the bizarre way we move around, which likely goes back to yet another evolutionary happenstance that ended up having major consequences. That is, it's possible that the earliest human ancestors developed a unique way of moving around that allowed for an upright posture to eventually become a biomechanical possibility. Fossils such as *Ardipithecus ramidus*—at 4.4 million years old, one of the candidates for earliest known human—suggest that some of our first fossil relatives were moving with unusual, upright postures while still in the trees. These dawn humans tottered around with a kind of awkward upright shuffle at least some of the time, moving above the branches rather than swinging below. When they started spending more time on the ground, then, they were already predisposed to walking upright rather than on all fours. Evolution carried that trend forward as other apes were shunted along different biomechanical paths. By the time of Lucy, however, our kind was already standing tall, and we can see that change in their vertebrae.

Even though we're often admonished to "stand up straight" as schoolchildren, the fact is that we can't ever truly do that. Our backs are curved. You don't even need to look at a skeleton to see this. Look at someone standing up from the side and you'll see that the spine takes a gentle S curve, sticking out at the shoulders and curving forward as it descends before curving out again as it meets the hip. This is strange for a biped. Dinosaurs like *Tyrannosaurus*, for example, were able to keep their spines horizontal. It helps to have a long, balancing tail. But for us, standing straight means having a vertically oriented, curved spine that evolved to help us balance our weight while running around. Part of that curve—where the lumbar vertebrae of the back curve forward—embodies what experts call lordosis, and paleoanthropologists are able to detect it even without a complete spine. This isn't a pathology. It's just the name for the inward curve low down in our spinal columns that's an important part of our biomechanical balancing act. Lumbar vertebrae that are part of those distinctive curves in our spines are wedge shaped—the height of the bone at the back decreases compared to the height at the front. Lucy and other australopithecines related to her had this kind of curvature to their spine, which, with other features in the skull, foot, and hips, indicates that they were walking in a way like we do. Their bones lock them into a particular way of standing and moving, which we can visualize even though we're millions of years too late to watch them move around in life.

Now, Lucy probably wasn't one of our direct ancestors. Drawing straight lines from living species to ancient ones along narrow paths of descent is extremely difficult, if not usually impossible, to do. Lucy fits in the same category of *Pikaia* and *Tiktaalik*—a transitional form that helps us understand how major anatomical changes transpired. Right now *Australopithecus afarensis* is as good a candidate as any for the ancestral stock from which some of the first members of our own genus, *Homo*, arose soon after, but the fossil record may yet provide us with an even better candidate. Nevertheless, the skeleton of Lucy and her kin reveal that by 3.3 million years ago humans had spines, hips, legs, and feet suited to striding around upright while they retained the flexible shoulders, long arms, and somewhat hooked hands that made their ancestors so adept in the arboreal realm just a few million years before.

We carry on that legacy, more than we often feel comfortable acknowledging. Lucy and her kind weren't all that different from us. There's a greater amount of skeletal difference between *Archaeopteryx* and birds fluttering around on the lawn or today's horses and their tiny, multi-toed ancestors like *Eohippus* than between us and

Australopithecus afarensis. Lucy was smaller than we are, her jaws jutted out further in front of her face, she had no tall forehead nor the expanded brain inside it, her rib cage was flared, and so on, but these differences are actually pretty minor in the context of broad evolutionary change. In other words, Lucy was unmistakably a human, even if her bones looked a little different from ours.

If only a skeleton from our species had been laid out next to Lucy's in that dark exhibit hall. Then the family resemblance would have struck home even harder. Not to mention that it might have given visitors a better appreciation for what's inside us. I know I didn't appreciate our own skeletal form until around the time I visited Lucy, when I was taking a human osteology course at Rutgers University. One cold autumn evening I showed up to the classroom early, only because it was better than waiting in the frigid air outside. A real human skeleton, origins unknown, stood next to the door on a metal stand, some of the bones labeled in fading black ink. I had never really looked at a human skeleton before. I'd walked by a few in museums, usually on my way to the fossil halls, and seen plenty of representations in cartoons and movies and other bits of pop culture, yet I had never stopped to admire the minimalist beauty of what our flesh conceals. We don't have any ornate decorations like *Stegosaurus*. And we're not big enough that our bones look hefty and overbuilt, like the giant ground sloth *Megatherium*. We're pretty simple as vertebrates go. Yet, even though our skeletons are boring compared to those of most other creatures, there's something wonderful about them. It was the elbow that got to me.

I'd never really thought about what's going on at my elbow joint before. I knew that the point is where my upper arm bone, the humerus, meets the two bones of my lower arm, the radius and ulna. And I knew that the "funny bone" that makes up the point of the elbow isn't a separate bone, and sadly isn't the humerus, but rather part of the ulna. But, up until that night, if you had asked me to sketch that joint I would have drawn a blank. I hadn't a clue. In my imagination, the bones just . . . met each other at a kind of unspecific hinge. It was only when I lifted up the skeleton's prone arm, humerus in one hand and lower arm in the other, that I started to understand. The end of the humerus flares out to the sides, with an odd spool-shaped protuberance on the front and a divot on the back. The ulna fits perfectly, its own end having a U-shaped articulation point that slides forward and back along the end of the humerus and, in life, is buffered by a cushion of cartilage. No metallurgist or machinist has ever made an articulation point that's as beautiful. And even better, while the radius doesn't do as much as the ulna in making the joint, its own end has a rounded, flat-topped mushroom cap that lets the radius swivel back and forth over the ulna. This is so simple, but it's been one of the most important factors in our success.

I'll bring in my fuzzy companions to help make my point here. My German shepherd, Jet, has forelimbs that are really great for the endurance running that dogs did back when they were still wolves, and they mostly move forward and back. If I hold out my hand, palm flat, and ask for his paw it's easy for him to slap his pads right on it. But if I hold my palm sideways, he can't slap or shake it that way. Not without moving his whole forelimb sideways or otherwise contorting to get his forelimbs into the right range of motion. My cat, Margarita, however, can easily accomplish the feat. Cats grapple their prey, whether that be a gazelle or a toy ball. Their forelimbs evolved to maintain flexibility and, like we can with our hands, turn their paws so they face each other or up to face themselves. That's all thanks to the way three bones come together. And you can have fun with this yourself as you're reading this. Hold out your arm straight in front of you, palm down. Now turn your palm up. Your radius is rolling over your ulna, which is barely moving at all. Go ahead and use your other hand to touch the tip of the ulna, your unfunny bone, as you rotate your palm up and down. It hardly moves at all, doesn't it? You can thank our tree-dwelling primate ancestors for that flexibility. It's a joint left over from a three-dimensional life in the trees co-opted to manipulating the world around us, a junction where two bones meet that dictates the range of motion available to us.

Our joints dictate the way we stand, run, grasp, sit, and every other activity that our bodies are capable of. It's easy to overlook this point. We molded the world around us—the tools and structures that form our lives—based on what our joints can or can't do. If you were shaped more like a deer, for example, trying to sit comfortably in a car seat would be an impossible challenge. The way we move, the way we live, is bounded by our bones.

Let me divert back to my beloved dinosaurs here for a moment. The standard impression of a *Velociraptor*—thanks in large part to the *Jurassic Park* films—is to fold in your pinkie and ring fingers, making the other three curved claws with your palms facing down. Very rawr. Only we know that dinosaurs couldn't do this, or at least they couldn't accomplish this pose in the same way. The wrists of raptors and other dinosaurs couldn't rotate like ours can. They were locked into a smaller range of motion that was more up and down—think of the simple, hingelike motion of a chicken wing. At rest, these dinosaurs kept the palms of their clawed hands facing each other and could achieve a palms-down position only by moving the rest of their arms (as some of the raptor-like dinosaurs did when they started taking to the air in the form of early birds). They were clappers, not slappers, and in an alternate history where highly intelligent dinosaurs ruled the earth instead of us, something as simple as a claw hammer wouldn't exist. The dinosaurs wouldn't be able to use it the way we do, instead probably relying on mouth and wing to fashion and use their own tools, just like some crows do today. The fact that our amazing hands are incredibly flexible appendages opened a world of possibilities to us that would be closed to other animals. Joints are what have opened and closed what we're capable of, and there's one more I especially want to draw attention to. It's so plain, such a constant part of our lives, that it's easy to overlook how truly strange it is.

With the exception of the sesamoid bones, strung along tendons, the 206 or so bones in our bodies rely on contact with one another. Our skulls are composed of various elements fused together into a single unit, while our spines are a swerving stack of individual vertebrae separated by discs of squishy cartilage, with our ribs attaching to their own vertebrae. Our hips are flexible but fairly simple. This is our ball-and-socket joint, with the head of each femur fitting neatly into a little crater on each side of the hip, while our knees are simple hinges. There's not much play for rotation since our legs evolved for the repetitive back-and-forth of walking wherever we desire to go. But the shoulder is the one that still confounds me. For as important as our arms are to us, in both the big picture of our history and day-to-day life, you'd think that our arms would be attached to our bodies with a really solid joint, something like a modified version of what's at our hips. Instead our arms seem to float on the outside of our skeletons. The skeleton in that college classroom, as well as the model of "Stan" beside my desk now, requires special nuts and bolts and struts to make sure the arms stay attached to the rest of the body. So how does it work when we're alive?

Your shoulder blades, or scapulae, are anatomically odd. They're triangular bones that sit over your back, gliding back and forth just behind your ribs. That placement, dictated during the days when our ancestors were still fish lashing their way through Devonian seas, is one of those evolutionary happenstances that has become critical to our history. It's a large part of why we can throw overhand, for example. If our shoulder blades were placed toward our sides, like on a dog or cat, we wouldn't be able to rotate our arms to hurl a spear at a mammoth or throw a fastball. We'd be stuck throwing underhand, like some baboons do when irate at safari tourists, and we'd mark the end of summer with the World Series of softball rather than baseball. That is, if we ever got around to inventing such a thing at all.

Where the shoulder attaches to the rest of the body, though, things start to get strange. One tip of the scapula opens up into a cup that receives the head of your upper arm bone, and just above that is a little flange of bone that connects to the edge of your collarbone, which in turn anchors to the central bone of your chest, the sternum. The whole apparatus looks incredibly flimsy. The whole set of bones, from the little flattened phalanges at the ends of your fingers up your arms to the shoulder blade, is really connected to the rest of the skeleton only by the tips of your collarbones resting at the base of your throat. Yet this flimsiness translates to flexibility and is part of what has allowed us to make our livings by manipulating the world around us. It's a gift from our ape ancestors, who couldn't have foreseen that monkeying around in the trees for millions of years would result in upper bodies incredibly adept at shaping the world to the whims and wants of their descendants.

What those early apes didn't have, however, was a foot suited for extended periods of walking upright. Their foot, so far as we can tell from the sparse human fossil record and our living ape relatives, was more like a hand, with curled toes and a sole that resembled a palm, the big toe jutting off to the side for opposable grasping. This allows great apes to walk upright for a time, but they have to shift their torsos from side to side to stay balanced and aren't especially adept at staying upright for very long. While our ancestors were not just like chimpanzees—chimps have been evolving for just as long as our human lineage has—the fossil trail has shown that the early, pre-Lucy human *Ardipithecus ramidus* had such a foot, and it would have required that "Ardi" amble around with a strange gait. Walking upright didn't mark the beginning of humanity. We started off in the trees, not so different from our closest ape relatives.

This was one of the greatest evolutionary trade-offs in our history, at least on the level of us all being preemies with squishy heads so that we might grow to have big brains as adults. Feet that were great at grasping branches and providing a firm grip for life in the trees just did not work for repetitive tromping around in the forest underbrush or out in the grasslands. The foot had to change. Hand-like dexterity was lost in favor of a foot with shortened toes directed straight out in rank and file, with the big toe brought into line with the others instead of jutting out to the side. We can wiggle our toes, sure, but compared to our hands, our feet are rigid structures that move on a simple hinge from front to back. Just try to move one of your feet to this side or that—much of the rest of your leg has to move because your ankle just can't contort that way. Our bones open possibilities and set firm limits.

Our evolution was constrained to give us such inflexible feet. Birds and the predatory dinosaurs they descended from, for example, required feet that could slash and catch and pin down food, and that's why the foot of a chicken in the backyard or a raven out in the desert isn't all that different from that of an *Allosaurus*. For our own lineage, though, moving out of the trees meant that there was no longer anything to grasp for support. A different foot shape was required to absorb the shock of step after step, not to mention a way for our feet to push off the ground at the end of each step and swing into position for the next. Doing this is the most natural thing in the world for most of us—like breathing. You don't need to think about each step as you amble along. But take a moment to get up and stroll slowly. Concentrate on the heel strike, the way the ball of your foot hits the ground and pivots your big toe into position to lift your foot off the ground as you balance on your other leg. It feels strange. But that basic weird-feeling motion is one of the most human things about us, and it's something we've been doing for at least 3.7 million years.

Even though anthropologists can, and often do, debate the fine details of the way our ancestors and relatives moved, a trackway discovered by anthropologist Mary Leakey in Tanzania provides indisputable evidence that humans were walking in much the same way that we do now way back in the Pliocene. That's because footprints, while not as sexy as bones and seeing the organism itself, are fossilized behavior. They are truly moments of time locked in the stone. And the Tanzania trackway, known as Laetoli, records the steps of at least three humans who walked over a pillowy bed of volcanic ash. Different researchers have interpreted them as a family, a couple followed by a child, a pair with a mother carrying a baby at her hip, or even unrelated people who passed by the same place within a broader window of time, and, frustratingly, there may never be a way to know any of that for certain. Out on a beach, for example, the footprints of various unrelated people cross and overlap and seem to follow each other yet were actually made at different times. All that can be said for certain is that the people of Laetoli were walking upright. And that's the key to our biomechanical story. There are no knuckle marks. No sign that these people were going down on all fours. The ash recorded the anatomy of a foot very much like our own, which probably belonged to a close relative of Lucy. By 3.7 million years ago the feet of early humans already had a big toe brought into line with the rest, allowing them to literally walk tall over the ancient landscape.

The ebb and flow of such major anatomical changes happens in the course of evolutionary time. Natural selection and other evolutionary forces were critical in making our shape. And, provided that we survive our self-destructive habits, they will continue to change us. We are still evolving. You can track the changes in our genes, as well as small tweaks to parts of our anatomy like our jaws and the microscopic structure of our bones as the nomadic lifestyle largely lost out to a sedentary and agricultural one. The way our ancestors moved, or didn't, has left their marks on our skeletons, too.

Though our bones are always shuffling around at the microscopic level, we can't change their shape through activity and exercise. We can artificially shape bones, as those many long-skulled cultures have done throughout history, and our bones certainly change shape as we grow up and then grow old, but no amount of moving and shaking can cause us to develop bigger bones or bones with vastly different anatomy. UFC fighter Ronda Rousey doesn't develop a different skeleton to match her physique before a fight. But this isn't to say that our daily motions don't change our bones at all. We can see the way the insides of our skeletons change with activity, and that's what allowed two groups of osteologists to determine that our species has changed with our more sedentary lifestyles.

Compared to our living and fossil relatives, we're pretty lightly built. We have what anthropologists call gracile skeletons—characterized by relatively low bone mass for our body size. That also means we're relatively fragile and more susceptible to bone diseases such as osteoporosis. And that's because our sedentary lives have changed our very bones. Two complementary studies published in 2015 highlight the shift. Using high-definition CT scans to focus on different parts of the body—one team looking at the head end of the femur, the other at sections of bone in seven different parts of our limbs near the joints—the researchers were looking for the density of trabecular bone. This is the specific type of bone tissue that forms struts and supports inside your skeleton, especially around joints that have to bear the brunt of our physical activity. In both studies, it turns out that sedentary, post-agricultural societies are literal lazy bones. Whether they were being compared to other primates, hominins, or even populations of *Homo sapiens*, people from sedentary cultures had lower trabecular bone density. People from more active, hunter-gatherer cultures, by contrast, had denser bones, with internal bone structure closer to what would be expected for a nonhuman primate of about the same body size. In fact, some fossil humans had more than twice the trabecular bone density we have today. They were more active, moving around the landscape and interacting with it in a much more physical way than we ever do. Even those of us who make a point of working out every day—or who make a living doing manual labor—still have less trabecular bone density and a higher risk of developing osteoporosis later in life thanks to this change.

It's possible the lower bone density in recent populations is a response to diet and daily activity, varying individually according to how much we move around. But, then again, maybe this is a real evolutionary change, humanity's near-total switch to a lifestyle centered around farming and diets based on grains having a lasting effect on the structure of our skeletons. Either way, we look pretty flimsy compared to the people who were trying to tame the Ice Age as they moved around the planet. And these sorts of mechanics don't only apply to our life on Earth. As we look elsewhere in the solar system for new places to explore, and possibly new homes, the way our bones react to the push and pull of daily life is a critical factor. If we can't solve the puzzle of keeping bones healthy in space, we're not getting much farther than the moon.

There are plenty of considerations if we're ever going to reach Mars with more than a rover: takeoffs, nutrition and comfort in flight, landing, survival, and more. But one thing that's not often discussed is what we're going to do about our bones. Bones thrive on physical activity. Stop moving and your bones know it. They start to resorb themselves, jettisoning calcium into the bloodstream and urine, not to mention making bones weak enough to incur a higher risk of fracture. We already know that weightlessness in space does this to astronauts. Space stations in orbit are basically in free fall as they go around our planet, and analysis of the occupants of *Mir*, for example, showed that they were losing 1 to 2 percent of their bone mass every month. And that's just the average. Some spacefarers have been recorded as losing as much as 20 percent of their bone mass during a six-month mission. That's a huge change, and it's going to be a problem for whoever wants to explore the red planet on foot. NASA estimates that it would take about nine months to get to Mars. Let's imagine our astronaut as they prepare to step out onto Martian soil for the first time. All those days stuffed inside a metal tube hurtling through the freezing void have led to this moment. They pull on their spacesuit, make sure everything's sealed up to protect them from the alien environment, and, excited, they skip the last rung of the lander ladder to hop down. Their first word on the surface might not be something poetic. It could be "*Ow!*" as they get a greenstick fracture on their weakened fibula.

There is at least one animal that gets around this problem, however. If NASA were able to train black bears for space exploration, they probably would.

Bears have the right idea about winter. They don't ski. They don't go out to shovel the drive. They do what I wish I could: hunker down in a sheltered spot to sleep through the chill. And given that we know bones are responsive tissues that need movement to keep replenishing themselves, we would expect the ursids to hobble out of their caves as frail creatures in desperate need of a workout after staying motionless so long. But this isn't what happens. The bears' body chemistry keeps their skeleton intact even as they snooze. In a 2015 study of thirteen hibernating black bears, biologist Meghan McGee-Lawrence and colleagues discovered that bears' bodies are able to slow down both bone formation and bone resorption as they hibernate. One of the keys is a protein called CART. The levels of this biomolecule in the bear's body shot up to as much as fifteen times the base level in hibernating bears, suppressing the amount of calcium from the bears' bones taken up by the bloodstream. At the same time, two other proteins involved in building new bone tissue—BSALP and TRACP—were reduced. This balance kept the bears' bodies in a new equilibrium. The bears become a closed system—no calcium intake or excretion during hibernation—perhaps giving us a model for how some future space travelers would be able to survive long journeys in low gravity. A bear bones approach may allow researchers to overcome the osteological issues of space travel.

In space or on our Earth home, though, bone is a reactive tissue that's shaped by what's around us. That's true both in the grand evolutionary sense—the modification of our skeletons giving us clues to how natural selection opened certain paths and closed others along our deep timeline—and in our day-to-day lives, the way we use our bones etching our history onto them. Our bones are time capsules of evolutionary and individual stories, but that's not all. The bumps, breaks, and diseases we suffer in our lives leave their own indelible marks on our skeletons.

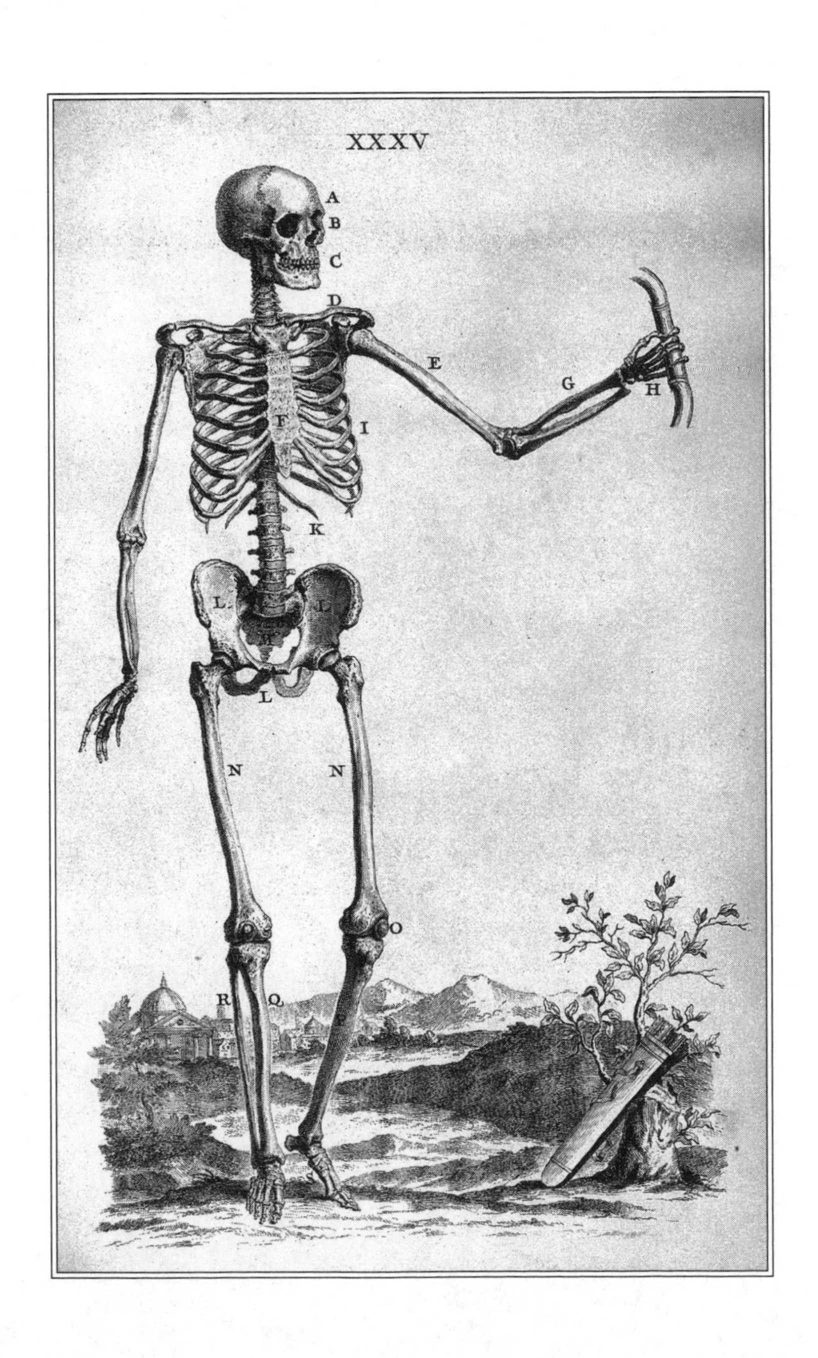
XXXV
A
B
C
D
E
G
H
F
I
K
L
L
M
L
N
N
O
R
Q
P
S

FIVE

STICKS AND STONES

Bone responds to the world just as other parts of our bodies do. When you cut your finger on a piece of paper (hopefully not the pages of this book!), you expect your platelets to plug up the exposed blood vessels and that your skin will eventually knit itself back together. Our bones can do much the same. When you break a bone, your body immediately begins a repair process to bring those two pieces back into accord. The way our bones grow and maintain themselves has provided us with a built-in repair system.

But sometimes bones don't behave as they ought to. Sometimes the elements that are supposed to be our internal supports become a kind of prison for the rest of our tissues, bending our bodies to the new course they've set. They are potent, if painful, reminders that skeletons tell the stories of our lives more powerfully than anything else.

Despite our intimacy with our own bones—there's a skeleton

inside you *right now*—it's all too easy to look at bones as objects. Anthropologists and anatomists might be able to look at this or that skull or other skeletal piece and say something about how old that person was or divine other clues from the osteological landmarks, but in broad strokes, bones taken out of their fleshy context can seem stripped of their stories. Except in the case of pathology.

Pathology is the study of the biologically unexpected. Most often it's concerned with disease and injury, but other alterations—such as the effects of corseting or skull reshaping that we encountered earlier—also fall under its purview, regardless of whether there were any detrimental effects to the person. In short, pathology compares bones to an idealized version of the complete human skeleton and notes anything that's different from that standard, with each of those anatomical departures also being called a pathology.

Pathologies are clues of a life lived, of bones broken and diseases suffered. We might not be able to get at the direct reasons for those injuries in every case, but they are nevertheless potent reminders that an individual was alive and still has stories to share. The bulge of bone on a healing rib or the faintest trace of a hairline crack on a piece of femur connects us with the dead more than a pristine skull does. Pathologies are touchstones that bring us closer to the life we're considering, raising questions that might never occur to us if the skeleton were in pristine condition.

Entire books could be filled with examples of pathological bones and skeletons. In fact, several have, and almost any human skeleton is going to show some sign of injury. No skeleton lacks signs of wear, even if it's just a tiny toe fracture you didn't know you had or a cavity that went unfilled. That's what allows pathology to connect us to people we'll never meet. The various imperfections our skeletons display are stories from our lives, privileged or long-suffering as they might be.

But while pathology may have principally been invented as a science by humans, for humans, it doesn't only apply to us. We're not the only creatures with skeletons, after all, and the way bone grows, breaks, and heals applies to other vertebrates as much as ourselves. In fact, the fossil record provides ample evidence that many of the osteological injuries we suffer aren't new at all. The list of bumps and breaks goes back millions and millions of years, each one adding a little more nuance to the stories of the creatures that capture our imagination in museum halls. In fact, sometimes the strange skeletons of ancient mammals and giant dinosaurs are so impressive that we miss these clues—tangible evidence of unique lives long lost.

One of my favorite examples stands in an alcove on the fourth floor of the American Museum of Natural History. It's usually quiet up there. The Milstein Hall of Advanced Mammals doesn't attract nearly the same crowds as the neighboring galleries of dinosaurs. But that's part of why the old *Megacerops* is always on my list when I stop by the museum. The skeleton of this beast—what looks like a towering rhino but belongs to an extinct, distantly related group of mammals called brontotheres—is ghostly white, the mineralization process that took place after its death some thirty-four million years ago giving it a pale shade approximating the colors of its living skeleton. It's beautiful. And if you look closely, the fifth rib on the right side looks a bit gnarly compared to its neighbors. About halfway down is a break surrounded by bulbous growths of bone from when the mammal healed. What exactly happened, no one can say. Maybe this *Megacerops* had a bad fall. Perhaps a rival slammed into its side and shattered the bone, just like quarreling bison do today. The information held in the skeleton doesn't go that far. But the ancient injury nevertheless records a painful moment in this individual's life and shows us that the animal survived it. The rib broke and was in the process of healing when something else killed the animal. And this small monument to prehistoric pain makes the old bones feel a little more intimate, easier to wrap up in imaginary flesh as we envision the reanimated bones stepping off their museum podium.

Of course, there's more to the skeleton than bones. Teeth present their own problems, too, with familiar dental ailments stretching surprisingly far into the past. Consider little *Labidosaurus*. This 275-million-year-old reptile looked like a medium-size lizard with a snaggletoothed overbite, and one particular specimen found in Baylor County, Texas, records the oldest known evidence of bacterial infection in a land-dwelling vertebrate. For some reason, perhaps biting off more than it could chew, the reptile broke off two of its teeth. Normally this wouldn't present much of a problem. These reptiles grew new replacement teeth throughout their lives. But in this case, bone covered the damaged tooth roots and trapped bacteria inside the jaw. The reptile suffered a severe bone infection, losing three more teeth and suffering an inflamed, painful injury that oozed pus. The degree to which its jaws were altered tells us that the reptile survived with this injury for a while, but every bite of each meal would have been excruciating.

And did you know that dinosaurs got arthritis? Many of us unfortunately become familiar with this generalized form of joint pain as we age, but fossil evidence shows that even the "terrible lizards" coped with aches that are familiar to us today. The way bone reacts to the loss of surrounding tissue tells the tale. While there are many, many different forms of arthritis, it generally manifests itself when the cushioning cartilage of a joint is worn or eaten away and bones now touch each other, new bone growth taking on gnarly shapes at the site of contact. It's one of the prices we pay for living to old age, but there are other ways it can happen. Sometimes an open wound might provide bacteria with a direct route to a skeletal joint, the microorganisms eroding the cartilage and generating pus as they make themselves comfortable, and one of the grossest examples comes from a pair of lower limb bones from a shovel-beaked dinosaur found in the sandy, sixty-six-million-year-old marl of New Jersey. The two bones, found fused together, were said by paleontologist Jennifer Anné and colleagues to have a "cauliflower-like appearance" toward the end where the two bones met at the elbow; the paleontologists describing the fossil called it "a mess of 'frothy' bone." This dinosaur's bone tissue was dying, the limbs rapidly growing new bone to try to make up for it, manifesting as septic arthritis that had eroded away the cushioning cartilage at the joint. In other words, this *hurt*, and the degree of bone growth shows that the dinosaur had lived with this problem for a long time before it eventually perished at the close of the Cretaceous.

Part of the reason we're able to recognize these problems in fossil creatures is because we know them in ourselves. It's the uniformitarianism of pathology. Our skeletons react much the same way to the same insults and diseases that vertebrates have had to deal with since the origin of bone. Having a hardened, internal skeleton has its drawbacks, and pathologies throughout history show us the natural risks that come built-in with bone. It's worth stopping to consider these various changes because they are unexpected expressions of what our skeletons can do, what they endure, and how they can repair themselves.

Pathologies also provide us greater insight into what our close hominin relatives and ancestors were doing throughout prehistory. While tooth cavities were once thought to be a relatively modern problem, associated with the invention of agriculture and greater reliance on starchy foods, a roughly fifteen-thousand-year-old archaeological site in Morocco has revealed that the hunter-gatherers buried there had really terrible teeth that were pocked with holes. About half of the adult skeletons found buried there had cavities. Acorns and pine nuts found at the same site likely explain why. These people loved their botanical sweets, and so their teeth look pretty similar to those of modern-day soda fiends. And, knowing what these cavities do to us, the researchers who described the discovery concluded that these hunter-gatherers "would have suffered from frequent toothache and bad breath," particularly since they lived about a thousand years before our species' earliest attempts at dentistry.

But it's not just bad habits that have become fossilized. The pathology of the human fossil record also suggests that we've been caring for each other for a very long time. KNM-ER 1808 is a celebrity skull among paleoanthropologists, a 1.7-million-year-old *Homo erectus* skeleton found in Koobi Fora, Kenya. But something was wrong with this person. The bones, identified as osteologically female, were not what researchers expected for a typical *Homo erectus*. Lesions marked her skull and jaws, and there were signs that her periosteum—that membrane that wraps around living bone—reacted terribly to some kind of pathology, her bones themselves showing signs of bleeding before death. These clues led researchers to suggest that she suffered from hypervitaminosis A. As you might guess from the name, it's a condition that arises from ingesting way too much vitamin A from sources such as fish or, as anthropologists suspected, the liver of a carnivore like a lion or hyena (although a second opinion suggests that eating bee broods could cause a similar vitamin A spike). Regardless of the immediate cause, this person was clearly ill for a long time—long enough for painful pathologies to accumulate across her skeleton—and she likely would not have survived as long without the help and care of others. Rather than being the brutish louts they're so often characterized as, early humans recognized illness and helped keep each other alive.

This is the history we've inherited—skeletons that record not only our evolutionary history and individual biology but also the way we lived and the pains we suffered. Osteopathology covers afflictions including bacterial infection, arthritis, and syphilis, certainly, but even something as common as a cavity or a minor fracture falls under the purview of the field. I have one such example myself. When I was ten, I borrowed an old, narrow skateboard from my grandparents' house. I was having a great time rolling down the hill of my family's driveway, seated, until my mother suggested that I try standing up on it. I promptly fell off and jammed my hand onto the ground hard enough that I got a greenstick fracture, a crack in the bone that didn't go all the way through, on my radius. I have a lot of memories from that summer of floating in the pool, cast-covered arm wrapped in a garbage bag. Thankfully the same processes that grow and maintain my bones eventually knitted the break together and likely obliterated all evidence that it was ever there, but I still have the X-ray to remind me of my minor misadventure. But it could have been worse. If I'd suffered a complete fracture—the bone actually snapping into two or more pieces—then the healing process would have taken far longer and required that the broken pieces be watched more closely to make sure they rejoined in the right way. And without medical attention, such breaks don't always knit back together neatly. Bone will still attempt to bridge complete breaks, bringing the ends back together, but they might be out of alignment, connected by messy growths of bone that alter mobility. In some extreme cases, the bones try to reach out to each other with new tissue but never actually meet, creating a false joint called pseudarthrosis. It's difficult not to cringe when looking at such injuries, but all the same, they speak to how versatile and even accommodating our skeletons are.

As we've seen from our prehistoric friends, though, breaks and obvious signs of trauma aren't the only starting points for pathology. Disease, nutrition, and other external causes have their role to play, too. If you go too long without sufficient vitamin C, for example, your bone tissue will start to thin and be more likely to break. That's one of the many terrible signs of scurvy and why it's wise to squeeze some lime in your rum. Likewise, persistent vitamin D deficiency prevents the osteoid of bones from properly mineralizing, so bones wind up being more pliable than they should be. That's why children who suffer from rickets often have legs that bow inward or outward. These are but two of the many ways what we encounter in life can alter who we are inside.

And then there's an entire category of what we inflict upon ourselves, intentionally or not. The way people through time and space have modified their skull shape counts as pathology, even if it doesn't affect health, as do the conditions of people who have altered their skeletal structure through corseting, amputation, and surgery. Fashion can alter our skeletons just as injuries can. So can the demands of music, as eighteenth-century Italian opera singer Gaspare Pacchierotti was intimately familiar with. While his origins are unknown, Pacchierotti made a name for himself as a fantastic mezzo-soprano, the removal of his testicles prior to puberty allowing him to develop a unique singing voice; he was a castrato. But the physiological effects of the alteration to his anatomy didn't just reside in the soft tissues of his body. He is thought to have been in his early eighties at death, and in addition to other pathologies—such as dental erosion that shows he was a persistent tooth grinder—there are open sutures on the singer's hip bones, which normally fuse and are obliterated by the time men reach their midthirties. This isn't surprising given that Pacchierotti essentially didn't get to go through puberty—his castration kept his skeleton, as well as his voice, in an immature state, and osteological meeting points that usually fuse did not. The same might be true of two much more ancient skeletons uncovered together in Quesna, Egypt. The two skeletons, which seem awfully tall for their estimated adolescent age, still have unfused parts of their skeletons that typically would have knitted together. While there are other possible causes, archaeologist Scott Haddow and colleagues suggest that the unusual traits might indicate these two people were eunuchs, the growth trajectories of their skeletons forever changed by what had been taken from them.

So far, however, we've been looking at the way skeletons react to the surrounding world—whether it be a nasty fall or the literal trappings of style. But there are also pathologies that find their origin from the inside out, arising from genetic or developmental differences that lead bones to deviate from the expected. Bones can be transformed in harmful, even fatal ways by mutations that affect the way our bodies grow. If a genetic alteration leads the pituitary gland of the brain to pump out too much somatotropic hormone, this spurs extra growth that manifests as gigantism. On the other hand, problems with cartilage formation during early growth can lead to dwarfism. Mutations affecting the formation of collagen in bones can cause a condition doctors call osteogenesis imperfecta—"glass bone disease"—which makes skeletons brittle. Alterations to the way our bones grow can have dramatic effects on the shape of our skeletons, sometimes not visible to us until a visit to the doctor for a broken bone or odd bump reveals something unexpected. That's what happened to one of the Mütter Museum's most famous residents, Harry Eastlack.

What's left of Eastlack is prominently displayed in a glass case on the bottom floor of the museum, safeguarding his bones while giving visitors a complete view of what happened to him. And unlike the skulls upstairs and some of the other specimens accumulated during a time when medical ethics were more questionable, Eastlack asked for this end. He wanted people to learn from what he had suffered. His skeleton, leaning to one side, head angled downward, is wrapped in struts and sheets of gnarly bone. It's as if a second skeleton started to grow over the first, wrapping up Eastlack in an internal prison. The degree of this transformation can take a moment to sink in, for your eyes to truly focus on and comprehend what you're seeing. As orthopedic doctor Frederick Kaplan observed, "Normal skeletons collapse into piles of loose bones when the connective tissues that join bones together in life are removed. To be displayed in human form, skeletons have to be re-articulated or pieced back together with fine wires and glue. As a result of the bridges, plates and ribbons of heterotopic bone . . . Harry's skeleton is almost completely fused into one contiguous piece."

Physicians call this disease fibrodysplasia ossificans progressiva, or FOP. It's comparatively rare, affecting about one in every two million people, but it's debilitating. And it underscores just how plastic our bodies are. When we think of bone, we think of all the disparate elements that make up our skeletons. They are discrete entities that fit together in a certain arrangement. The leg bone is connected to the hip bone . . . But bone is also a tissue that, under certain circumstances, can form in places where it shouldn't be. In cases of people afflicted with the heritable disease FOP, soft parts of their bodies become transformed into bones. Ligaments, muscles, and other soft tissues ossify into bone—creating struts and bridges that start to constrain the body from within, what Kaplan calls "an armament-like encasement of bone."

Doctors have known about the disease since at least the eighteenth century. John Freke, a London surgeon, noted that on April 14, 1736, a fourteen-year-old boy came to ask what could be done about the swellings on his back that he had suffered from since he was a small child. Upon examination, Freke found, "They arise from all the vertebrae of the neck, and reach down to the os sacrum," or essentially all along the spine. "They likewise arise from every rib of his body, and joining together in all parts of his back, as the ramifications of coral do, they make, as it were, a fixed bony pair of bodice."

Even centuries after Freke's initial diagnosis, the disease is hard to spot. The majority of cases are initially misdiagnosed as signs of tumors, bunions, or other problems. Children born with the disease often have malformations of the big toe, but even then, some people don't know they have FOP until bone starts to show up in the wrong place. In the case of a twenty-one-year-old woman who went to the hospital complaining of shortness of breath, for example, a radiograph revealed that one of the muscles of her back and nearby soft tissues had started to turn to bone, constricting her rib cage. (And humans aren't the only animals to be afflicted—FOP has also been reported by veterinarians in cats and dogs.) It doesn't take much to kick off the transformation. Small injuries and trauma set off a cascade of ossification, usually starting with tissues on the back side of the body. That's how it happened for Eastlack.

Eastlack was born in November 1933, but no one knew anything was wrong until years later. When he was five years old, while playing with his sister Helene, he was struck by a car and left with a broken leg. The fracture didn't set right, and shortly thereafter the boy's whole leg seemed to stiffen up. When doctors had another look they found bone was growing, but in the wrong place—within the muscles of Eastlack's thighs. Even then, however, doctors didn't know they were dealing with FOP or that continued operations—like trying to remove the excess bone—were actually making the condition worse. The ossification was relentless. Little by little, growths of bone across Eastlack's body locked his skeleton together so that he could barely walk even with the help of a cane and his jaw was fused into motionlessness. He was almost forty when he died, stating in his last wishes that he wanted his skeleton to be displayed so that people could learn about his disease and that perhaps there would be some secret inside that would lead to a cure for others so afflicted. To date, there's little to do for those who have the disease other than identify it and try to minimize injuries, "a reality so stark, sobering, and inescapable," Kaplan writes, "that it transcends the imagination." But Eastlack remains as a source of information, a monument, and a person. "When an important FOP discovery is made," Kaplan says of the research team studying the disease, "we return to visit Harry's skeleton in order to confirm the physical and the anthropological reality of the discovery."

What happened to Eastlack is a powerful example of how tenacious bone can be. Bone has many ways of catching us by surprise and showing up unexpectedly. We expect soft tissue to turn into bone while we're embryos and infants—the growth and fusion that locks our skeletons into a biomechanical apparatus able to protect our organs and let our muscles move us around—but such transformations can happen much later in life. Remember that Geza Uirmeny, upstairs from Eastlack at the Mütter, was surprised to find that part of his larynx had turned to bone. This is not the most cringe-worthy place that unexpected bone may grow.

We don't have genital bones. The *os penis* and *os clitoridis* of other primates were totally lost during the evolution of our early primate ancestors. Other mammals still have them, from cats and bats to monkeys and rabbits. But we don't. Other primates do, but our anthropoid ancestors lost them somewhere along the evolutionary trail. So that's why it might seem a little weird that some men find themselves suffering from penis bones.

In case you're crossing your legs right now, dear reader, at least I can say that this isn't a particularly widespread problem. Only about forty cases have ever been reported. But even though it's something of another anatomical outlier, why it happens can tell us a little something about the strange nature of bone. There's no single cause for the problem. True, the root cause—as with almost any other part of the body—is bone cells working to transform what was once soft tissue. But there's a whole range of reasons why this metamorphosis might happen, ranging from kidney problems and STIs to trauma. In 1933, for example, a doctor noted a nineteen-year-old man who had a mass of bone in the tip of his penis. "The patient also had a gunshot wound at that site 3 months previously," a medical review later reported as the likely cause of the pathology, although what must have been the fascinating story of the initial accident was apparently not recorded. And even though some physicians toyed with the idea that such conditions were an evolutionary throwback—to when our monkey-like ancestors had literal boners—the fact is that it's another example of how bodies can make bones all over, even in places that have no connection to the skeleton, from breasts to salivary glands.

And here, we pivot again. Bones react to the way we live, and the details of our own biology can sometimes cause bone to form in unexpected ways. But there is a third class of curiosity that pathologists concern themselves with. These are the marks we've left on bones both living and dead, the history of our osteological understanding literally etched onto our skeletons. Consider trepanation.

Brain surgery didn't wait for the days of anesthetic and delicate tools. If anything, it's an incredibly old discipline that humans have been trying their hand at for thousands of years. The record is in the skulls. From Europe to the South Pacific, from Africa to the Americas, people invented and practiced this remedy with varying degrees of success. The oldest example so far reported comes from Sudan. While excavating a roughly seven-thousand-year-old settlement, archaeologists uncovered the grave of an old osteological male. The body lay curled up on itself, in the fetal position, and on the body's skull the experts found a nickel-size hole where the bone had been carefully scraped away. It's the most ancient example yet found of this intentional modification. The question we're left with is why. The bone tissue that had been removed doesn't show any signs of healing. If this person didn't die in the procedure, they likely perished soon after. Archaeologist Łukasz Maurycy Stanaszek also pointed out that the hole could have been made after death for other reasons, like giving a spirit inside the body an escape route. What is clear is that whoever performed this procedure was already practiced. The scrape is even and smooth around its borders, indicating that this wasn't the practitioner's first operation.

But the all-time experts on trepanation seem to have been in the Inca Empire of Peru. More than eight hundred examples of trepanation have been found from the Inca heyday between the fourteenth and sixteenth centuries AD, spread geographically all over the culture's span. In this case, anthropologists think they have a better explanation for why the practice was so popular. For the Inca, trepanation wasn't based on the idea that there were spirits that had to be released from the body or some other supernatural purpose. It's more likely that this was a form of surgery to remove shards of bone from fractured skulls. These people, anthropologist John Verano observed, used weapons that caused blunt force trauma, such as clubs and sling stones. Their weaponry didn't pierce and slash, but smashed and bashed. That led to some pretty gruesome head injuries, and so those with medical knowledge turned to trepanation to pick out the dead pieces of skull and create new, even holes that would gradually heal with new bone growth.

The procedure probably wasn't as painful as it sounds. The skin of the scalp is brimming with nerves and blood vessels—meaning that creating a window to the skull would have been an uncomfortable and bloody mess—but the bone of the skull itself doesn't have the same network. There's no reason for it to. Our skin is the barrier that tells us about the outside world. Bone is there to protect, having that distinctive layout of pliant, spongy bone sandwiched by denser protective bone plates. So getting through the flesh would have been far more painful than going through bone, even though hearing someone scratch a stone tool against one's skull must have been a unique experience.

The number of healed holes speaks for itself. About 83 percent of the holes the anthropologists examined had at least begun to heal, and signs of bone infection are minimal. Most patients survived the operation and the wounds stayed clean as they started to knit back together. More than that, the locations of the holes say something about just how much these Inca specialists knew about the skull. "Trepanations were placed in cranial regions that avoided musculature and other vulnerable areas of the skull," the anthropologists found, and it seems that the Inca surgeons also knew about the locations of the blood vessels pressed against the brain that should be carefully avoided. Granted, these methods had to be learned and passed down, and, like anything in human culture, the practice varied from place to place across the empire and over time. Even so, the evidence shows that the Inca surgeons were precise in the methods they chose, and the practice was a common medical remedy for what were apparently extremely common head traumas across their lands.

The Inca experts weren't the only doctors to leave an osteological trail of their surgical practice through time. In Western societies, sometimes archaeologists turn up relatively recent burials that show signs of dissection. A slice running all the way around the skull to access the brain for study is a good giveaway. And just like the Inca skulls, these bodies marked by Western medical practices can tell us something about how doctors operated. In 2015, Jenna Dittmar explained to the assembled American Association for the Advancement of Science meeting that the skeletal trail left in English hospital graveyards shows how human dissection practices changed between 1650 and 1900.

Understanding the story took some piecing together. The remains found in these cemeteries, Dittmar noted, are often isolated limbs or skulls. That's because a single body often had to be split between multiple anatomy students. During the nineteenth-century heyday of anatomical instruction, in particular, grave robbers did a brisk trade trying to supply the ever-hungry medical schools with cadavers, the demand for study corpses always outstripping supply. But whether acquired legally or by nefarious means, the remains indicate how medical instruction changed, cut marks from dissection tools giving away how students were breaking down bodies for study. At the beginning, the dissection tools weren't at all refined or specialized. They were more like carpentry tools than medical instruments. Over time, however, the cuts get smaller and the apparent violence done to the dissected bodies seems to decrease. That probably didn't matter much to the dead, but it made a world of difference to the patients who were operated on with some of the same tools.

Most of our tale so far has focused on the natural history of bone. We've traveled through the life of bone, from its evolutionary origin through its biology and the way it reacts to the world. But pathology can often speak to the inevitable connection between life and death, as a skeleton from Central America reminds us.

In 2017, archaeologist Nicole Smith-Guzmán and colleagues reported on the remains of an unfortunate teenager who had been afflicted with several pathologies. The skeleton, buried sometime after AD 1250 at Cerro Brujo in Panama, had pitted teeth, lesions on the skull associated with anemia, and bone cancer in the upper arm. The sarcoma created a bulge of painful-looking bone in the teenager's right humerus, and likely contributed to this person's early death. But even though they were buried in a trash pit, it's the context of the burial that really tells us something about who this adolescent was. This person was wrapped tightly and buried with care, interred with objects including a shell trumpet. Based on what was known about the local culture at the time, Smith-Guzmán and colleagues think that the bones represent someone imbued with special significance in relation to life and what comes after, the death growing in their bones making them the embodiment of a connection between the worlds of the living and the dead. That's the realm that we're entering now, where those who yet live try to make sense of the deceased. These are the nested stories of osteology—in other words, not just the evolutionary, biomechanical, and biological life of bones, but also where dead skeletons intersect with people who are still alive. How bones are viewed, and what we think they tell us, varies according to our perspective, cultural history modifying how we assess and understand natural history. When bones no longer speak for themselves, we speak for them.

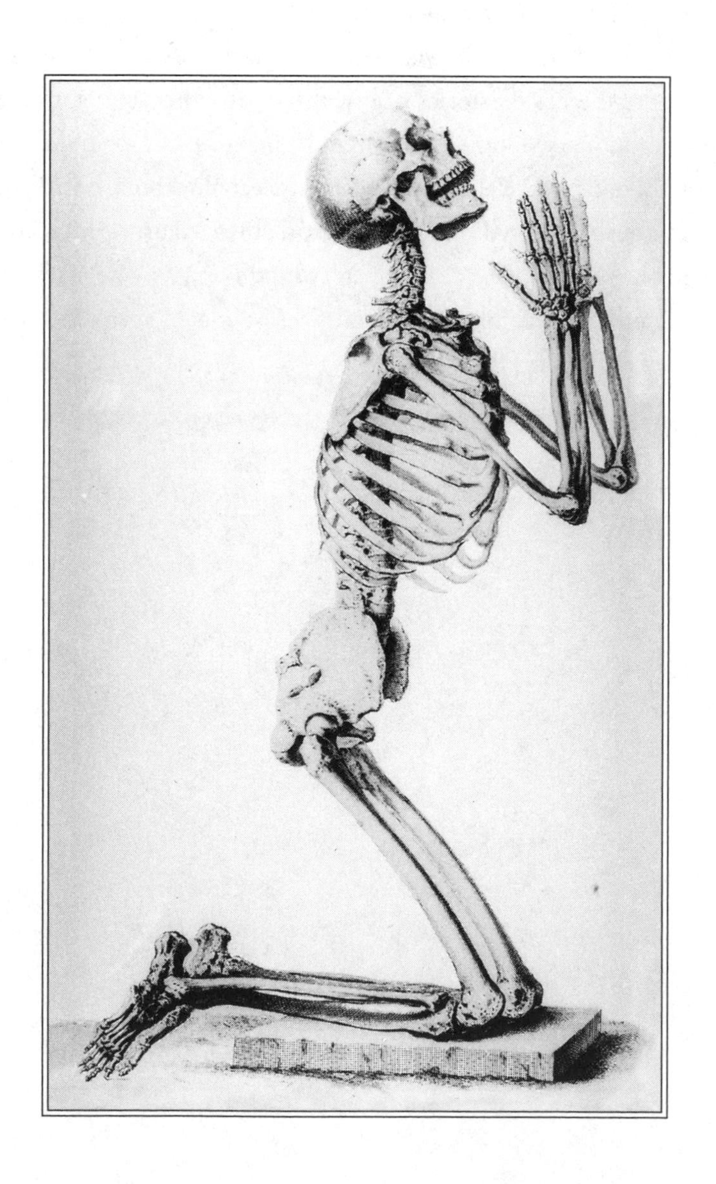

SIX

THE NEARER THE BONE, THE SWEETER THE MEAT

There are plenty of reasons to visit St. Bride's Church along London's Fleet Street. For one thing, the Anglican house of worship is one of the oldest in England. The architectural bones of the current structure date back to about 1672, but the church has existed in some form since the seventh century. The original church was turned to ash in 1665's Great Fire of London, but a new building was soon constructed on the spot. More than that, its tiered tower is rumored to have been the inspiration for a lovestruck baker to create something a little more extravagant than the traditional—and innuendo-ish—"bride's pie": a decadent, stacked cake that couples continue to shell out obscene amounts of money for. And as if that weren't enough, St. Bride's is the "journalist's church." Newspapers funded the restoration of the house of worship after it was bombed by the Luftwaffe in 1940, underwriting its current incarnation.

I didn't know about any of that history as I weaved past Fleet Street businessfolk, smartphones permanently attached to the sides of their heads like brain slugs, looking for the church entrance. It was a pleasant and sunny spring day, but all I wanted to do was get underground. That's because St. Bride's was the only place in London I could get to with skeletons in the basement.

The technical term for what lies beneath St. Bride's is an ossuary. It's a pleasant, sibilant word for what amounts to a bone-storage room filled with the dust of the dead. An ossuary may be a box, stacked skulls in an underground room, or something as ornate as human bones built into the very structure of the building. Each and every one has its own aesthetic. The Brno Ossuary beneath the Czech Republic's Church of St. James is an assortment of skulls, eye sockets gazing back at you in full attention. The Skull Chapel, beneath St. Bartholomew's in Poland, is far more ornate, the skulls in the ceiling undergirded by crisscrossing thigh bones, with walls made up entirely of seemingly innumerable crania, all gazing at Christ hanging from the cross atop an altar. And then there's the Sedlec Ossuary, also in the Czech Republic, where skulls are strung along the ceiling like popcorn lanyards on a Christmas tree, with dangling strings of skeletal elements creating the effect that you've just stepped into David Cronenberg's foyer.

These places might send a shiver down your still-living spine, and they certainly raise the question of how all those bones were acquired in the first place. Their backstory is far more prosaic than their ambiance: it's all about space. When you have people inhabiting a given patch of land for centuries upon centuries, graveyard elbow room is going to be at a premium. Bones last for a very long time, and though we're not exactly as prolific as rabbits, we reproduce faster than decay recycles the dead. So to make space for those newly arrived to a postmortem existence, churches from the Middle Ages through relatively recent times would exhume some bodies and stack those bones inside holy storage rooms. (One of the most recent, France's Douaumont ossuary, holds the remains of more than 130,000 unidentified soldiers from World War I's tragic Battle of Verdun and was established in 1932.) And if you're going to go to all that trouble of digging up, cleaning off, and reorganizing bodies, why not get a little creative with the arrangements? For many ossuaries, there is an art to the practical solution of cramming bones beneath the church; there is beauty in the dead.

I was hoping to see one of the more ornate storage rooms beneath St. Bride's, and I tried to avoid any photos that might spoil the experience as I did my initial research for my afternoon in the crypt. Understandably, a guided tour is required—having camera-toting visitors like me disturbing the deceased just won't do—so after paying the requisite £6 I tried to keep from blurting, "When do we get to see the bones?" as I patiently nodded and "Hmm"ed through most of the hour-and-a-half-long tour. I'm sure the church itself knows this, saving the subterranean charnel house for the very conclusion, and at long last our guide led us into a low, dim room where an anthropology researcher pored over old bones on a small desk in the corner. Even though this was an official Anglican church tour, it was the sort of space that made me half-expect a sharp *thwack* against the back of my skull before I joined those already held inside the ossuary. It's a room that makes you shiver, grave in both the figurative and literal senses. And through an open doorway, stacked on the floor, are the bones. There are no garlands of femora or pyramids of skulls arranged by monks who had a lot of time on their hands. Long bones sit piled in parallel rows, with a few skulls resting atop the ranks in the room of ancient charred brick. It's a gloomy place where dust easily catches in your throat, and it has the more utilitarian feel of a storage room rather than anyplace that was ever meant for display. It's not a haphazard collection. While it certainly makes the most of the space, it lacks the artistry of other ossuaries.

What makes the remains here special is that the bones of St. Bride's were actually better kept than some of their counterparts. For their initial interment, parishioners were buried by the church in sturdy lead coffins with their names, date of their demise, and cause of death. This has allowed association between the bones of St. Bride's and the backgrounds of the people those elements once belonged to, offering demographic and historic information that otherwise would have been lost. Had the bones been turned to art, then they'd be inaccessible—stripped of context as they were of flesh. Mosaics and garlands made from remains are beautiful, sure, but putting those pieces back together and trying to verify who someone was is nearly impossible. The more practical approach of St. Bride's ended up providing anthropologists and archaeologists with a way to check ideas drawn from bones against demographic data to better identify remains and what they can tell us. It's a way of groundtruthing. Anatomists have compared St. Bride's bones with historic information to double-check that osteological sex really can be determined from hip bones, for example, and a more recent study tried to determine whether using the end of the fourth rib to estimate age at death is reliable. (It's not, as it turns out, and so experts need to turn to other methods.) But this combination of bones and historical information can also give us some insight into the lives and concerns of people who died hundreds of years ago. For instance, there are more records of infant deaths at St. Bride's during the time the cemetery and crypt were active in the eighteenth and nineteenth centuries than are actually found inside them. The reason for the discrepancy, archaeologists suspect, is that the cost of burial in lead coffins was prohibitive and families who had lost many children simply wouldn't be able to afford the church's standard practice.

But bones aren't just for scientists or for tourists like me to gawk at. What a bone is, and what it means, depends on who you ask. Where I might look at a skull and see the visage of a slightly modified ape, hurtling along a continuing evolutionary story, an anthropologist might see a representation of a particular time or culture, a pathologist might see abnormalities, a collector might see a curio, the pious might see a saint, or anyone might see one of their ancestors. These lenses are not mutually exclusive, and we often switch between them, imbuing bones with meanings and qualities that go beyond the biological details wrapped up in our osteology.

Given that bones are the most lasting parts of us, it's only natural that they have complicated afterlives. Just look at how we preserve them. Bones are naturally at home in bodies, of course, and throughout human history we've deposited the deceased in the ground. Not every culture does this (the Maasai of eastern Africa, for example, traditionally leave their dead exposed to be recycled by scavengers), but burial goes back a long way, and isn't even unique to modern humans. Neanderthals have emerged as a tender example of the thought and care people put into the disposition of the dead.

Neanderthals have suffered a long-running PR problem. They were characterized very early on in the nineteenth century as thuggish half-apes, inferior to our own ancestors labeled as *Homo sapiens sapiens*—they had stockier builds suited to hunting Ice Age game, heavy brow ridges, and lower foreheads, all of which contributed to a popular image of these people as musclebound brutes. The division between us and them was drawn starkly, the fact that Neanderthals went extinct seemingly being the only proof needed that there was something wrong with them. We had art and invention and culture while the Neanderthals were the stereotypical fur-clad cavemen who thought about little more than meat. More recently, even as Neanderthals have developed a rehabilitated image that includes that realization that they are the same species as us, we still look down on them as somehow deficient or culturally sluggish. After all, there's no one alive today with the traditional Neanderthal osteological build, and their culture apparently disappeared. They lost out to us, even as fragments of their biology live on in our genes. (DNA research has left no doubt that people from our culture and Neanderthals swapped genetic material, meaning that many of us carry on the Neanderthal legacy through a handful of surviving genes that continue to be passed on.) But, piece by piece, the story is changing. For a long time Neanderthals were thought to be symbolically inept, lacking anything that can be called art. We now know that's not true—new analysis of cave paintings in Spain has shown they were made by Neanderthals, not modern humans. And from the way Neanderthals buried their dead, they must have had a deeper understanding of life and what came after than we've traditionally been willing to admit.

For decades, archaeologists were resistant to the idea that Neanderthals intentionally inhumed their dead. The bones and skeletons found in caves could have been covered by rock falls. But as experts looked closer, investigating new sites as well as refining dates from old ones, they found that chance alone couldn't explain sites where multiple Neanderthals had come to rest thousands of years apart. These burials were intentional, and it stretched credulity to think that stone tools, bones from other animals, and feathers all wound up in the graves just by luck every single time. Once the dumb Neanderthal bias was peeled away, the real people started to emerge.

The fact that at least some Neanderthals buried their dead with forethought and care offers a hazy, but significant, look at their minds and their culture. "It is difficult to imagine that a human group could excavate a grave, position the corpse in the pit, and offer funerary goods with no form of verbal exchange," Francesco d'Errico and colleagues pointed out. At a roughly 60,000-year-old site in Amud Cave, in Israel, archaeologists found a burial of a 10-month-old infant with a red deer jaw over its hip. In the 75,000–45,000-year-old sediment of Syria's Dederiyeh Cave, there was a burial of a 2-year-old with a chunk of limestone placed near its head and a piece of flint over its rib cage. Then there's La Ferrassie, France. Here, archaeologists have found at least eight Neanderthals. Most are very young—from fetal to ten years old—with an adult woman and adult man nearby, and the majority of them seem to be associated with some sort of grave goods in the form of bone shards and pieces of stone tool. From Uzbekistan to Iraq, from France to Israel, groups of Neanderthals intentionally buried their dead over thousands of years. Their care for the dead is why we have their skeletons to study at all.

What Neanderthals thought about death, and about bones, we can't know. Given the diversity of our own interpretations, it would be wrong to project any one set of beliefs onto them, much less modern ones. (In fact, this happened to a skeleton known as Shanidar IV—pollen of plants with medicinal properties were found around the bones, leading to the conclusion that the plants were intentionally laid there and these were the original flower people, but later analysis indicated that the wealth of pollen was likely spread around by local insects and other pollinators.) All the same, the way Neanderthals treated their dead shows that humans—even ones outside what we may think of as our strict species—have been thinking about death for at least tens of thousands of years, adding symbolism to the natural reality. And considering our evolving interest, it was only a matter of time before we started giving a face to Death.

There's never been one manifestation of Death. The visage I know is the one from heavy metal album covers, Halloween costume shops, and Terry Pratchett novels—a scythe-carrying skeleton robed in black. But that's just a symptom of the time and culture I live in. Over and over again, people have reinvented Death not so much as a force but as a personality, both good and bad.

The ancient Greeks had Thanatos, a god of death who was just doing his job. Someone had to bring the living to the land of the dead, and the winged deity was the one to do it. The Aztecs, by contrast, had Mictecacihuatl, who was less concerned with transport between worlds and instead watched over the remains of the dead.

Supernatural beings like these kept the transition from life to death orderly. They are not evil but simply carry out cosmic duties. But especially in the wake of the Black Plague, which ravaged Eurasia during the fourteenth century, manifestations of death took on a more sinister turn. This is when Death, as a figure, became frightening. The loss of life was so great that it permanently seared itself into the cultural consciousness. During this time the image of Pesta, the plague hag who went from town to town deciding who would live and who would die, spread through Scandinavia. The classic image of Death as the Grim Reaper emerged in the wake of this fourteenth-century European catastrophe, as well, the cloaked skeletal frame reminding us what we would become given enough time, the scythe reaping us from this world and into the next. But despite this common embodiment of our fears, interpretations of how Death is supposed to behave vary. Sometimes Death actually kills the person who is set to die and takes them. Other times Death is more of a guide and emissary, showing up for those who have already perished. Either way, the image stuck in the public consciousness, and now Death is recognizable most anywhere you go. Skeletons became synonymous with the grim and ultimate end we face, and that has only fed our enduring fascination.

We've had a long and complicated relationship with skeletons of all kinds, dating back to when early humans cracked open long bones to get at the marrow inside. While perhaps not as ancient as cut marks from the time of *Homo erectus*, there's cultural evidence that we've been transfixed by bones for a very long time, as demonstrated by the legends prescientific cultures told to make sense of fossil bones. Yes, a skull or skeleton is a memento mori of what awaits us, but that's not all. Skeletons support the symbolism of our cultures. Take, for example, what archaeologists have called the world's earliest skull cult.

Göbekli Tepe, situated in southeast Turkey, is considered to be the oldest temple yet known. The ruins of the ten-thousand-year-old place of worship are circular, with the remains of pillars and walls still standing and looking as if the roof had been sheared off. Thanks to analyses of bones recovered from the site during years of excavation and published in 2017, archaeologist Julia Gresky and colleagues found that many of the 408 skull fragments recovered from this place had cuts or holes in them. These weren't accidentally made by rough treatment after burial. They were intentionally made by people, cut marks showing where flesh had been removed from the bone and at least one of the shards showing a hole that had been drilled into it so that it could be suspended by a cord.

No one knows who the skulls belonged to. There are no human burials at the temple. There are just the fragments. It's impossible to say what the intent of the skull collectors was, whether these were people being honored or defaced. But this was a place where skulls and their component pieces were treated with reverence and interest, and that's why the people who left these traces behind have been referred to as a skull cult. And they weren't the only people to be so fixated. Various skull cults through time have developed their own beliefs about skulls and what significance they hold to the living and the dead. At a site in Syria called Tell Qarassa North, for example, the faces of skulls are cut and mutilated—what archaeologists have interpreted as ritual punishment in the afterlife. But whoever was in charge of Göbekli Tepe had a definite interest in skulls. In addition to the fragments, archaeologists found cultural clues that underscore a peculiar sort of osteological obsession. There are figurines with their heads removed, small sculptures of heads being presented as gifts, and other artwork depicting decapitated individuals.

Of course, a fascination with bones isn't just an ancient activity. Bones are still important to the faithful, even if they aren't always front and center of a religion. Relics remain. Buddhist, Hindu, Muslim, and Christian cultures have all collected and venerated remains thought to retain special properties after death. But no one does it with quite as much panache as Catholics. Something as simple as an unidentified bone fragment may be enshrined in ornate trappings of fine cloths and precious metals that the living saints probably would have eschewed in life. And while the remains themselves do not have the power to heal, the belief goes, they are a connection for those coming to have an audience with the remains and God. It's a signal boost for parishioners looking for intercession.

Wish fulfillment isn't the only aspect of relics, though. Part of their appeal, author Peter Manseau writes in his book *Rag and Bone*, is that these remains are presented as not just a *what* but a *who*—someone of special significance to the faithful, a more macabre version of briefly spotting a celebrity dining in the same L.A.-area restaurant as you. They become celebrities among the faithful, connections to them after they pass on. And they can still draw crowds. In early 2018 the Jesuits of Canada and Catholic Christian Outreach toured the severed arm of Saint Francis Xavier to cities across Canada. As far as arms go, it's a fairly famous one. Francis Xavier was a sixteenth-century missionary who was supposedly the first to visit Japan and Borneo, baptizing thousands of people along the way. He also established the barbaric Goa Inquisition, which punished, executed, and burned Hindus who had converted to Catholicism and subsequently returned to their previous religion. This terrifying legacy didn't seem to bother the event's founders—"We love him to bits," said Catholic Christian Outreach cofounder Angèle Regnier during the tour—and the severed arm, which was supposedly Francis Xavier's baptizing arm, drew thousands of worshippers at each stop before eventually returning to Rome. And this isn't the only purported piece of the saint around. Legend has it that in 1614 one of his toe bones was bitten off by a visitor to his corpse while it was on display in Goa, India. Exactly what possessed the woman to do so isn't known, if the story is even true, but according to lore, the toe was eventually recovered and the bone is still at a church in Goa, while other parts of his body—such as parts of his arms and internal organs—were later more ceremoniously divvied up among the faithful institutions who wanted a piece of him.

But relics are a tricky business, sometimes literally. Fakes and frauds abound, and the bones of venerated saints are not always what they seem. The bones of Saint Rosalia, a Christian hermit who died in 1160, are venerated in Sicily in a shrine located in Palermo. This is despite the fact that nineteenth-century naturalist William Buckland identified them as goat bones in 1825, and they have since been removed from public view. Other fraudulent relics were intentional scams. Ever since people started creating relics from the holy dead, there's been a brisk trade in parts of the venerated. "I've photographed at least six different St. Valentines," relic-focused photographer Paul Koudounaris says. And lest this seem strange to you, keep in mind that this is a religious expression of what's an apparently irrepressible urge to own something from the influential and famous. You can bid for "authentic" Elvis Presley hair locks on eBay, and a UK-based collectibles shop offers a supposed snippet of Marilyn Monroe's hair, cut moments before she sang "Happy Birthday, Mr. President" to JFK, the secular equivalents of the religious relic business.

Still, bones don't necessarily need celebrity to transfix us. Skulls, in particular, have long been treated as collectible items. Those fused sets of bones are the most powerful skeletal expression of self, holding our personality even in death. A femur or a vertebra feels impersonal—like it could be from anybody—compared to a skull. And even though people have been creative enough to make calendars and flutes out of long bones, a great deal of the record of human bone modification centers around the skull. Skull cups, in particular, have been all the rage for quite some time.

"The use of human braincases as drinking cups and containers has extensive historic and ethnographic documentation," anthropologist Silvia Bello and her colleagues write, "but archaeological examples are extremely rare." That makes the bones found in Gough's Cave, England, very special. Within the confines of the cave, anthropologists recovered thirty-seven cranial fragments from human skulls. About fourteen of those could be fit back together, and the anthropologists suspect that the bones represent at least five people—a three-year-old child, two adolescents, and two adults. The bones themselves tell their postmortem fate. They're hazed with cut and chopping marks from when their flesh was removed. This was no random operation. The cut marks replay how the people who lived here detached and skinned the heads. Even in the dry language of science, the description is chilling: "Cut-marks on the areas of insertion of neck muscles and the presence of cut-marks in proximity to the foramen magnum indicate that the head was detached from the body at the base of the skull." From there, the muscles attaching the lower jaw to the skull were cut, the masseter and temporalis muscles winnowed off, with the tongue, lips, nose, ears, cheeks, and eyes taken off in the process. This was not some murderous frenzy. This was careful, deliberate work. And once all those soft tissues were removed, the skulls were struck to remove the facial bones from the dome of the cranium, taking off the face to leave the bowl of the skull intact. The people who did this didn't let useful tissue go to waste. The lower jaws were broken in such a way to expose the marrow inside, just as these people did to the jaws of horses, deer, and lynx found in the same cave. Cannibalism might have been part of the process.

The grisly Gough's Cave cups were made about 14,700 years ago. They weren't the only ones of their kind. Two other sites in France of comparable age have a preponderance of skull fragments that were altered in a similar way. Other people hit on the same idea throughout time. I guess there's just something about a skull that makes some people want to drink from it. Neolithic remains from Herxheim, Germany, and Bronze Age finds at El Mirador Cave, Spain, demonstrate the popularity of skull cups over time, and histories from those written by Herotodus to China's *Records of the Great Historian* to Mágnus Ólafsson's *Krakumal* mention people who quaffed from the crania of their rivals, with the historic, ritual use of skull cups attributed to Australian aborigines, people of Fiji, some Indian religious sects, and more. It's never really gone out of style.

Of course, manufacturing skull cups has been just one tradition among many. What skulls can be used for, and what they mean, is apparently limited only by human imagination. A pair of skulls found at a Neolithic site in 'Ain Ghazal, Jordan, were originally thought to have cut marks on them. But a reinterpretation by archaeologist Michelle Bonogofsky found that the scratches on these were from a different kind of bone modification—sanding and the addition of plaster to model the skulls after death. Other people tried their hand at skull decoration, too. Painted or otherwise modified adult skulls from the Neolithic have been found at multiple sites off the east coast of the Mediterranean—an area called the Levant. One skull found in Jericho had cowrie shells in place of the eyes and applications of plaster, including an artificial chin to replace the missing mandible. Another, from a site called Beisamoun in northern Israel, "was coated by a slab of plaster in the form of a horseshoe that was bound to the mandible and provided a realistic view of the modelled face," the plaster painted with rust-colored pigment. Whoever KNH-Homo 1 was appears to have been special. This skull, found at Kfar HaHoresh, had a detailed plaster mask complete with mouth, eyes, and cheeks modeled in plaster, perhaps an attempt by the artist to capture something about what this person looked like in life. And as different as each modeled skull looked from each other, archaeologist Yuval Goren and colleagues found that each was made in its own peculiar way. There wasn't any one single tradition or right way to modify these skulls. The plasters used to decorate them were made with different mixes—even for skulls at the same site—and fillers used for the eyes and other parts of the skulls depended on the sediment nearby. What each of these works was supposed to mean is unknown. Each of the skulls seems to be from someone young. Nor is it clear whether there was someone in each community with the specific purpose of making these works. But it's still an outgrowth of our obsession with skulls and the symbolism of where life and death meet.

This fascination hasn't withered. It'd be easy to think of skull cups and ritual decapitation as parts of a barbaric past that we've left behind us. But the power of bones is still there, whether it's to honor the dead, embarrass our enemies, mock death, shock us, or simply make a buck. In the collections of New York's Metropolitan Museum of Art, for example, there's a strange instrument with a mysterious past. It's a human skull retaining a ring of hair and festooned with antelope horns, with gut strings stretched over the open cranium. The artifact looks like a set piece for *This Is Spinal Tap*, and it's easy to believe that such an instrument could be used to play the dead into the next world or riff on the reputation of slain opponents. But the truth of the matter is probably much more sinister than the defleshed skull would initially suggest. The instrument-like object was purchased from an unknown dealer somewhere in Africa during the nineteenth century. No one has ever seen its like. No known culture makes instruments like this. Other people have made music with bones—the museum also holds a Tibetan drum made out of human skull bones—but no one has made anything as metal as the skull lyre. The most likely answer is that some con artist just made the instrument to sell to a European looking for something unique. It wouldn't be the first time.

Back across the Atlantic, in a case huddled among the exhibits of Oxford's Pitt Rivers Museum and labeled "Treatment of Dead Enemies," there's a display of several shrunken heads. These were made a century ago by the Shuar of Ecuador and Peru, and to the unenlightened visitor they would seem to be a sign of these people's ferocity. Few would notice that not all are human—there's a sloth in the mix—or, as anthropologist Frances Larson relates in her book *Severed*, how these heads came to the Pitt Rivers. The heyday of Shuar headhunting was fueled not by any personal animus, but by the market demand of the shrunken heads they made. Europeans wanted shrunken heads for curios and museum pieces, and the Shaur wanted guns and other resources; commerce was the motivation behind the phenomenon, instead of the traditional and relatively rare practice of taking heads for the power they were believed to contain. "When visitors come to see the shrunken heads at the Pitt Rivers," Larson writes, "what they are really seeing is a story of the white man's gun."

Thoughts of skull cup processing, cannibalism, and decapitation might all feel very uncouth and barbaric to us, just as it did to European and American anthropologists. But we don't need to look back in time or to abused cultures to find examples of bones treated badly. And when it came to acquiring skulls, headhunters weren't some far-off people. They were the supposedly enlightened European class, as in the craniology craze of the late eighteenth century. We had more faith in bones then, even if it was woefully misplaced, and the details of a person's skull were supposed to reflect the genius—or lack thereof—within. This is where obsession and science intersected, the skulls of the famous being both status symbols and potential clues for understanding the origin of genius. So aspiring craniologists would steal the skulls of the noteworthy to investigate if there was an osteological secret to their heroes' greatness. The skull of doctor and founding skeptic Thomas Browne was one of these ill-gotten prizes.

Browne feared what might become of him after death. We know because he put his fears on paper in 1658, before the craniometry fad took hold. His worries had more to do with being turned into an object. "To be gnawed out of our graves, to have our skulls made drinking bowls, and our bones turned into pipes to delight and sport our enemies, are tragical abominations." It's probably for the best that he never found out what happened to him in the centuries that followed. After a peaceful postmortem rest of more than a century, Browne's skull was stolen in 1840 when grave diggers accidentally uncovered his last resting place as they worked in an adjacent plot. His skull was removed to make a cast for craniometrists to go over at their leisure—even though it didn't include Browne's facial bones—but when the replica was finished the church grounds-keeper, George Potter, had other plans for the skull. He knew such a famous head was worth more than its weight in gold—skulls tend to be light, anyhow—and after a bit of angling about he sold Browne's most characteristic bones to the doctor Edward Lubbock, who left them to the Norwich and Norfolk Hospital Museum. When the vicar of St. Peter Mancroft petitioned to have the skull returned in 1893, the museum didn't budge. Working from that cherished precedent of finders keepers, they argued that no one could have a legal claim to Browne's head and so they owned it. It wasn't until 1921, almost three centuries after Browne's death, that his head was re-united with his body. Funny how skulls can lead some of us to become lobotomized on matters of ethics.

And celebrity still draws us to particular skeletons, whether they're scientific symbols—like the fossils of Lucy I just had to see—or the remains of the noteworthy and powerful. Fame and reputation seem to seep into the nooks and crannies of bones themselves, turning our gaze as we look to the past. They become symbols of their time and place, standing out in front of their contemporaries as connections to times we can never experience. Skeletons can be celebrities, too.

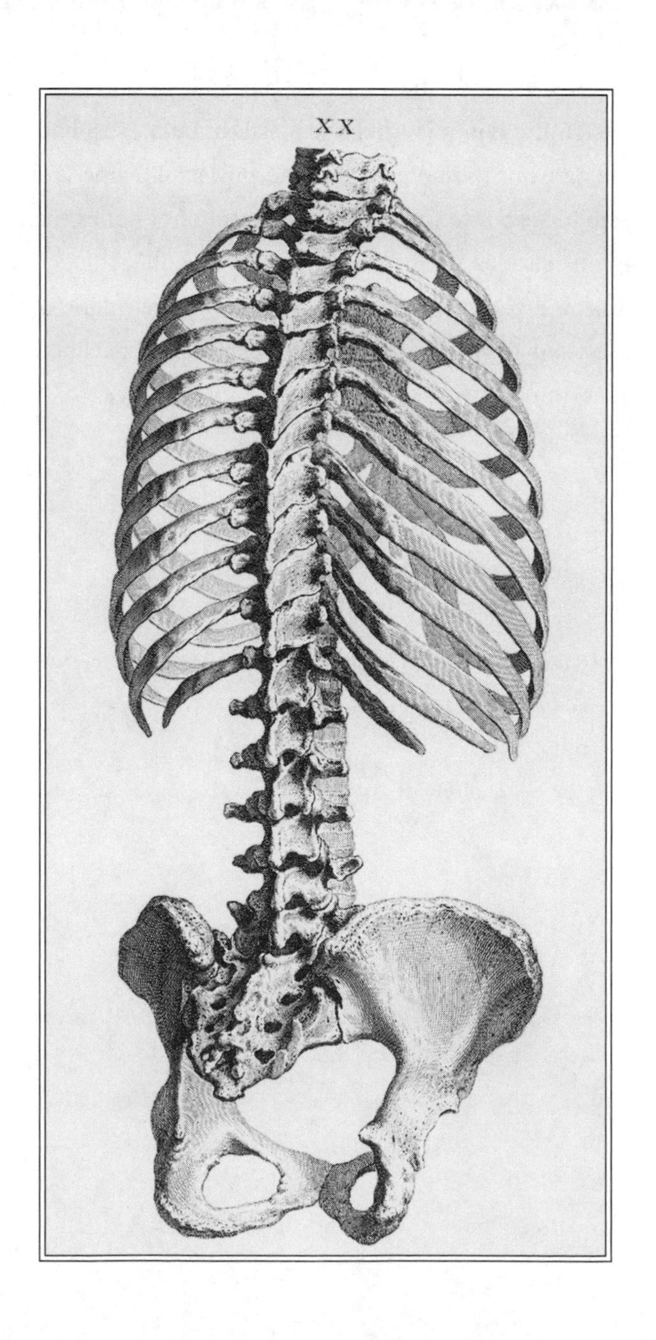
XX

SEVEN

BAD TO THE BONE

I never met Richard III in high school. *Hamlet* was the English department's prescription from the Bard, and honestly, even if we had been assigned the more dramatic Shakespearean tale about the hunchbacked ruler, I don't think I would have remembered much. I was always the sort of kid who would pick up Peter Benchley before anything suggested from the classics section. But in 2015, a series of dramatic headlines piqued my interest in the English ruler. After centuries being presumed lost forever, Richard III was shaken from his six-century slumber beneath a parking lot.

Such identifications are not taken lightly, and figuring out who's who from a pile of old bones is no easy task. There are only a handful of bones that can definitively indicate someone's biological sex, and bones are often mute—or at least incredibly confusing—when it comes to details like ethnicity and other attributes that we're quick to assign when looking at someone in the flesh. But still, anthropologists do their best to draw secrets from what remains.

How skeletons are uncovered, collected, and studied varies from case to case. Who is doing the excavating, where, and the nature of the site all make a difference. The way paleontologists uncovered the person inhumed at La Brea differed from the excavation of the crypt beneath St. Bride's, for example, and forensic analysis of a possible crime scene is going to proceed differently than the study of a historic burial site. Still, where data collection is concerned, archaeologists and anthropologists work according to certain practices to make sure that they're more or less on the same page; documenting osteological sex, estimating age at death, inventorying teeth, and investigating alterations to skeletons and bones after death are all textbook stuff.

But outside of the nitty-gritty of standards and practices, it's the questions about the past that drive the science. Writing for *Forbes*, bioarchaeologist Kristina Killgrove shared what it was like to study a Roman villa called Oplontis that was torched by Mt. Vesuvius in AD 79. It's as good an example as any of how modern experts go about their work, from the kinds of questions that direct the research to the methods used.

Given that Oplontis was a historic spot—and not a mystery burial out in the boonies somewhere—there was already plenty of background on the villa and who might have lived there. But the real draw was the Oplontis skeletal collection. Previous excavations in the 1980s had turned up more than fifty skeletons crammed together against the walls of one room. Some were removed; others were left exactly where they've rested for centuries upon centuries.

With her graduate student Andrea Acosta, Killgrove started going through the boxes of removed skeletons with the goal of reassembling some of these poor individuals. "We measure them, look for evidence of disease or healed fractures, and painstakingly catalog their dental health," Killgrove writes, and that's because, as we know, pathologies and other signs of life can offer clues about who these people were and what they were doing so long ago. As for the skeletons still in place, Killgrove and her colleagues scanned and photographed them before removing them for study, creating a record of the room as it was when they arrived. This is the bread and butter of bioarchaeology, carefully cataloging and analyzing bones in order to say something about the conditions under which people lived and died.

Naturally, not all skeletons turn up where you expect them to be. Archaeologists expect and even anticipate finding human remains while poring over the remnants of historic lodgings or cemeteries, but skeletons also have a habit of popping up unannounced. Sometime in May 2015 a strong windstorm ripped through County Sligo, Ireland. The powerful gusts upended a thick tree that had been growing for more than two centuries, and the upended roots were wrapped around the top half of a skeleton. Upon examination, the remains turned out to have belonged to a roughly twenty-year-old osteological male who had apparently been killed by a blade more than a thousand years ago—knife wounds marked the bones of his hands and ribs, and it appeared he was given a standard Christian burial for the time.

The same basic questions about the bodies in Oplontis and the tree-tangled skeleton in Ireland—who they were, what their lives were like—surrounded a mystery skeleton found beneath a British parking lot. This was the skeleton that would eventually be revealed to be Richard III, but that conclusion would require various lines of evidence drawing on new technologies such as CT scans and isotopic analysis as well as comparative anatomy. It's challenging to piece together anybody's life from their remains, but the demands of archaeology are even greater when you're claiming to have found a British monarch last seen in 1485. As royalty, poor Richard got an amount of scientific and analytical attention possible for just about anybody but usually reserved for particularly intriguing cases. There are plenty of bodies that could introduce us to the process by which a mystery body is identified—new ones turn up practically every day—but in terms of celebrity, scrutiny, and the unexpectedly bizarre ways some people react to the hoary dead, it's hard to beat Richard III. It's something of a reversal from Emerson's adage "When you strike at a king, you must kill him." If you're going to exhume someone's past, you may as well pick a king.

Richard III did not have a regal burial. He was not interred in a castle or any grave dripping with the status of aristocracy. The dead resting in your local graveyard received more ceremony and ornamentation than Richard got in the end. The long-lost king wasn't even given the luxury of a headstone. At the time of his discovery he was nothing more than a body grinning up at his discoverers from a hole in the ground, not a scrap of apparel, jewelry, or other artifice to identify who he was. Recognizing the identity of those bones was an exercise in high-stakes anthropological detective work. You don't want to claim you have a king when all you possess are the bones of some poor medieval schlub. Only his bones could testify to his identity and his life, and if he had been found at any other time in history it's likely his fragile remains would have moldered away on a shelf as another osteological unknown.

But before we can get to know the real Richard, we need to visit his specter, the one that Shakespeare conjured up to menace the English crown.

Shakespeare's famous tragedy set the tenor for much of Richard III's legacy. The aristocrat was a "poisonous bunchback'd toad," the "slander of thy mother's heavy womb." With everyone throwing such invective at him, it's no wonder the play's Richard leans into it and says, "I am determined to prove a villain / And hate the idle pleasures of these days." And, befitting such unmitigated venom, different actors have depicted English history's most despised ruler with varying levels of camp. Laurence Olivier was the canonical celluloid Richard III. He comes off as a menacing knave, hobbling around and mugging toward the viewer as if they're a coconspirator in his plans. From the very start, there's no question that he's the villain, letting you in on his murderous schemes of taking the crown for himself right at the outset. Among other machinations, this involved having his nephews murdered in the Tower of London. As classic as Olivier's rendition of Richard III is, though, a more recent characterization makes a better touchstone for comparison to the man who was so recently found beneath a British parking lot. I'm talking about Benedict Cumberbatch's portrayal in 2016 in two episodes of the TV serial *The Hollow Crown*. Just as the real Richard III was being introduced to the public through science, Cumberbatch's dramatic take more than doubled down on the ruler's bad reputation.

With Cumberbatch heaping on the menace, *The Hollow Crown*'s

Richard III bears a prominent hump on his back and, as the play dictated, a withered arm. It's driven Cumberbatch's Richard to madness. The haughtiness of Olivier's version is replaced by an unhinged, growling charlatan who only strays further into darkness as the drama progresses. In the end, as the play and history dictate, he meets his comeuppance at the Battle of Bosworth, impaled by a spear before collapsing into the mud.

There are some, however, who don't care for these dramatic characterizations at all. Centuries after his death, the slain king has several fan clubs who insist that history—and Shakespeare—got him wrong. They say the play was nothing but part of a smear campaign to malign the reputation of a man who was, they contend, a just ruler. They call themselves the Ricardians.

The Ricardians are not exactly unbiased observers of history, focused on uncovering the truth come what may. They've been on a crusade to exonerate their favorite king and clear his name from Shakespearean slander. Imagine that five hundred years from now, the grave of Michael Jackson has somehow been lost and most of what people remember of him is the *Thriller* video. A diehard team of Jacksonians, then, might become obsessed with relocating his remains to prove once and for all that the king of pop was not a werewolf. It's that kind of impulse that led Richard III superfan Philippa Langley to team up with University of Leicester Archaeological Services, an archaeology firm housed at the college, to search an area that seemed like just the right age and location to potentially hold some sign of the much-maligned king and proof that would clear his name.

No one was expecting that the lost king would turn up. History was clear that after a short, two-year reign, Richard III was slain at the Battle of Bosworth, the finale of the Wars of the Roses, on August 22, 1485. Not wanting to let a good PR stunt pass by, Henry Tudor—soon to be crowned King Henry VII—flopped Richard's body over a horse and paraded the deceased king to Leicester. After three days, Richard's naked body was taken down and buried at a place called Greyfriars. From here, though, the story gets hazier. There was a report that Henry VII put an alabaster tomb over the grave, but his successor, Henry VIII, had the buildings destroyed in 1538 after the building had already lain in disrepair for years. The burial site was lost. A later report from 1611, by contrast, suggested that Richard III's bones were tossed into the River Soar, but there was never any evidence to corroborate this. The most anyone knew was that the king was initially buried at a place called Greyfriars and was likely to still be there. Enter Langley. She urged the archaeological project to relocate the Greyfriars church in hopes that Richard III would be safely ensconced inside.

For Langley, what Richard III was like in life was not a mystery. She already had her image of him set in her mind. He couldn't have been the hunchbacked devil that Tudor propaganda had turned him into. His real bones, if found, would reveal the truth. And so, in a rare collaboration with amateurs led by Langley, the ULAS researchers sifted through lore and old maps to try to locate the spot where the nearly mythical Greyfriars church might be. The archaeologists were able to pinpoint the general area, but for the exact disposition of the church they would have to dig.

The most probable location of Greyfriars was narrowed down to a city council social services parking lot. It looked as spectacular as that name suggests—just asphalt and the ass ends of cars crammed together. ULAS archaeologist Richard Buckley, who had already been working on the project, had initially surveyed the area with ground-penetrating radar in hopes of finding some sign below. The results were inconclusive, but the project went ahead anyway.

Given that the archaeologists were uncertain exactly where the church might be, they decided to excavate long exploratory trenches to see if some signs of the building might pop up out of the old soil. Even then, the ULAS crew was not expecting to find much beneath the asphalt. "Owing to the lack of good local building stone," a report on the dig later pointed out, "medieval walls in Leicester are normally found to have been extensively robbed of material from both superstructure and foundations, and floors rarely survive in good condition." Whatever was left of the church and its contents would add a little more history to the area, but anything the researchers uncovered was likely to be an archaeological mess given the ravages of time.

No one expected to find any bodies, much less royalty. The excavators, although funded by the Ricardians, put locating Richard III dead last on a list of goals for uncovering the friary. "Archaeologists today do not as a rule seek to excavate the remains of famous people and historical events," Buckley and his collaborators later wrote of the project. That's partly because a king actually isn't all that informative about life in his own time. His position of privilege skews the information that can be read from his bones. The skeleton of a laborer with bad teeth would yield more about the basic details of fifteenth-century existence. All the same, the very prospect that a king might be found beneath the asphalt drew out re-enactors as well as the media. Even risking a letdown like Geraldo Rivera's embarrassing unveiling of Al Capone's vault, the reporters and camera crews didn't want to miss anything. If things went bust, they could still very easily package up a neat story with a cliffhanger ending: "The search continues . . ."

But, as luck would have it, the dig was on the mark. Remnants of the cloistral buildings, the church, and even several graves all turned up—intriguing stuff for aficionados of old England between the thirteenth and sixteenth centuries, providing new evidence about a town center that stood for three centuries before being demolished. But it was a nearly complete, peculiar skeleton that had everyone talking. In the very first trench, in what used to be part of the choir, were the bones of an osteological male. (The shape of his hip bones gave away that much, at least.) As archaeologists began to brush away sediment from the spine, though, this skeleton stood out as more than just a random burial. The spine took a tortuous turn, immediately recognizable as advanced scoliosis. Richard III was famous for his pathological back. Could this be Richard III? The prone skeleton, laid out as if just taking a six-century nap, demanded an answer.

"The body appears to have been placed in the grave with minimal reverence," Buckley and his colleagues later wrote, inhumed in an "untidy lozenge" in the ground. There was no evidence of a coffin or even a shroud, and the way the skeleton seemed to slump against the side wall suggested that the body was simply lowered into the grave and not even re-centered for a more aesthetic burial. This was basically one step short of dumping him in a random ditch. That itself could have been a clue—Richard III's burial was said to have lacked the pomp and circumstance that we might associate with a king—but it wasn't yet enough to make a positive identification. The body was in the right place, at the right time, to be the missing king, down to a real ailment that could have inspired legends of a hunchback, but all of these factors could have simply been coincidences.

There are various ways to begin to read a skeleton. Naturally, there are the bones themselves. Age, stature, pathologies, osteological sex, and other details can readily be ascertained by looking at anatomical landmarks of the skeleton that are well known to anthropologists. But there is more to it than that. There's the historic age of the skeleton and the context in which it was found. If there are objects or artifacts nearby, these might narrow down that person's culture—the people they associated with. And then there are secrets held within the bones—the genetic, chemical, and microscopic clues that can help experts focus on who someone was and how they lived. It would be this smaller set of clues that would eventually confirm the tantalizing prospect that the lost king had finally been found. It was genes, not gross anatomy, that brought us face-to-face with Richard III.

Bones are living tissue, and it's only natural that cells that help constitute bone contain DNA. But for someone who's been dead as long as Richard III, the process isn't as simple as scraping off a bit of bone and popping out a whole genome. For starters, DNA starts to degrade at death. In fact, as examination of extinct flightless birds called moas has shown, DNA degrades according to a half-life, breaking down into tatters until nothing is left. The maximum window for DNA preservation after death is a little more than six million years—not even close to getting us *Jurassic Park*'s dino DNA but more than adequate to give us a sample of the last Plantagenet king.

When humans dig anything up, though, we run the risk of contamination. Saliva carried by a breath, skin cells scratched off on bone, or other sources might wind up masking the actual DNA of the person being studied. Archaeologists have to cover up and be careful when excavating and storing old bones so as not to compromise the sample. It might be more than a little alarming to run a DNA analysis on someone long dead and come up with a readout that says it was actually you buried in that hole.

Rather than trying to reconstruct Richard III's complete genome, which would be impossible with the passage of time, geneticist Turi King and colleagues focused on a particular sort of genetic clue: mitochondrial DNA. This is the series of A's, T's, G's, and C's held inside a specific organelle—the mitochondria, forever immortalized as the "powerhouse of the cell" in Biology 101 classes—that is passed down through matrilineal routes. There was another advantage to this. Even though Richard III has living relatives from the patrilineal side, King and colleagues pointed out that "the male line is far more susceptible to false-paternity than the female line is to false-maternity events." At least with matrilineal analysis, the most confounding factor was surname changes rather than the potential for secret liaisons that never made it into the history books. And seeing that Richard has no living descendants, close relatives had to do. With a bit of historical background research, two living female-line relatives of the dead king were found and their genes were compared to what was assembled from Richard. Michael Ibsen, nineteen generations removed, and Wendy Duldig, twenty-one generations removed, are fourteenth cousins of each other twice removed, but in their genetic makeup were the details that would allow Richard III to be confidently identified. The skeleton really was the missing king.

The experts even came up with some extras. DNA isn't just about our relatedness to one another. Recovered genes can also tell us something about what people looked like in life. In the case of Richard III, this allowed a check on artistic depictions of the king.

There are no existing portraits made of Richard III in life. The oldest painting of the ruler was made about twenty-five years after his death. But in the course of their genetic analysis, King and colleagues put in the extra effort to look for markers of hair and eye color. Richard III seemed pretty plain in his portraits—it's not as if he had fiery red hair and ice blue eyes, which would be relatively rare—but it's still good to double-check. The eye color seemed to match the blue in the portrait, and while the hair results came back as blond, the geneticists pointed out that these results usually reflect childhood hair color and many blond kids grow up to be brown-haired adults. What Richard's face actually looked like was lost when his flesh sloughed away in the grave, but the tattered DNA in his bones confirmed that the medieval painters got at least some details of his appearance right.

Chemical traces had even more to say about Richard. The critical clues are chemical isotopes that enter the body and end up being locked inside teeth and bones as they grow. These markers have been used to track how whales went from landlubbers to full-time sea dwellers by looking at oxygen isotopes that tracked their intake of seawater, or to examine what kind of prey saber-toothed cats preferred by comparing carbon isotope signatures with those of the herbivorous animals they lived alongside. We're animals, too, of course, and some of the same clues can help us at least narrow down aspects of a person's life, such as what they were eating.

For many dead vertebrates, like prehistoric reptiles, teeth are often the most sought-after reservoir of these informative isotopes. As teeth form, isotopes from intakes such as drinking water and food get bound up inside them. The problem is that we're mammals. If we were reptiles and continually formed replacement teeth during our life, then isotopes from a tooth would offer us clues relatively close to the time of death. But most mammals only form two sets of teeth—milk teeth and adult teeth. That means that isotopes are only getting preserved in our teeth during our childhood. Richard III's teeth can say something about his diet during early life, but they won't help us reconstitute his last meal before his death at Bosworth. So, given that bones can be reservoirs of isotopic information as well, the researchers working on Richard's skeleton were able to fact-check where the ruler was raised and the general menu he picked from when he was an adult.

The researchers looked at geochemical isotopes relating to a different aspect of what Richard was ingesting. Water sources fed by rainfall, such as rivers and springs, for example, will have different chemical signatures of the oxygen isotope $\delta^{18}O$, and this isotope is taken up in our skeletons as our bodies grow and maintain themselves. The carbon isotope $\delta^{13}C$ does the same, only instead being tied to food sources such as particular plants, the animals who ate those plants, or the carnivores who ate those herbivores. Together, these different profiles allow experts to do things like track how diet changes when someone moves to a new place or as they grow up. Working from samples from Richard III's femur, rib, and teeth, the team attempted to reconstruct an outline of the king's lifestyle.

Historical sources say that Richard was born in Northamptonshire. The strontium isotopes—which are affected by the geology of where food is produced—indeed matched eastern England. On top of that, differences between the chemical signatures in Richard's bones highlighted his rise to power. Bones are always growing, and over time they replace themselves. A femur takes about ten years to completely turn over, for example, while a rib takes about two to five years. All those busy little osteoblasts and osteocytes are constantly reforming our skeletons bite by bite. That means the bones the researchers sampled offered two different snapshots of Richard's life, with the values from the femur representing a longer average of his adult life and the values from the rib tracking much more closely to what he was dining on when he was king. So when the researchers found higher oxygen and nitrogen isotope values in Richard's rib as compared with his femur, they took this as a sign of an influx of fresh fish and waterfowl in his diet, a trend for the wealthy. He was enjoying luxury foods later in his life, even if dishes like pike and egret sound more strange than fancy now. They also found something that at first seemed odd. The oxygen isotope values, which are tied to water intake, seemed to hint that Richard had lived in a place with a lot of rainfall. This didn't match the historical records. But then the experts realized that wine probably holds the answer. Kings drank wine like water, and the way oxygen isotopes change in making consumables, such as brewing beer, fermenting wine, or stewing food, probably caused the discrepancy between where Richard was and what his skeleton suggests. Overall, they concluded, Richard's bones show "a significant increase in feasting and wine consumption in his later years." If only we all were so lucky.

Despite these luxuries, though, even kings suffer pain. In Richard's case, it was in his back.

Thanks to Shakespeare, Richard III has always been portrayed as a hunchback. But *hunchback* has no real medical meaning. There's more than one condition that can make someone appear stooped over, and the bones of the king demonstrate that he didn't have the grotesque hump that *The Hollow Crown*'s makeup team placed on Benedict Cumberbatch. In fact, historian John Rous wrote in 1490 that Richard III was short-statured and his right shoulder was higher than his left. And this is what the king's skeleton showed when it happened to turn up at Greyfriars. It was obvious right away. Richard's spine had a swerve right in the middle rather than the rigid, straight appearance of an anatomically healthy vertebral column. Just to be sure, archaeologist Jo Appleby and colleagues did a CT scan of the spine and reassembled a polymer replica of the king's back. What they found confirmed Rous's report and the initial suspicions of the archaeologists: Richard had scoliosis. What caused the condition? The spine doesn't show the anatomical markers of it being present from birth. Nor are there signs that it was related to cerebral palsy or other conditions scoliosis can be associated with. Instead, the researchers describe the scoliosis as "subtle," indicating that it happened during the last years of Richard's growth spurt around age ten.

As extreme as Richard's spine looks, though, it would be hard to tell from the outside. Appleby and colleagues write that he had a "well balanced curve." So his trunk would have been relatively short compared to his arms, and his shoulders would have been at different heights, but it wasn't so noticeable that it couldn't be covered up. "A good tailor and custom-made armour could have minimized the visual impact of this." There's no sign that Richard hobbled, limped, or had scoliosis so extreme that it could have impaired his breathing. So much for Olivier's hobbling, capering depiction.

But the strangest thing about Richard III's skeleton wasn't his unceremonious burial or his contorted back. It was what had been done to him on the battlefield. Let's just say that Cumberbatch's depiction of the king got off easy when the end finally came.

Richard III died in battle; there's no doubt about that. But how did he perish? As anthropologists studied his bones, they found an unexpected amount of trauma. All together there were nine injuries to various parts of his skeleton, none of them healed. These were made just before, during, and just after the king's death.

Being able to read bones for battle damage is a relatively new science. Despite the fact we've been hacking away at each other since we invented blades, and archaeologists have been curious about battlefields for as long as the discipline has existed, very little work has been done on figuring out what kinds of weapons create what kinds of injuries. We know more about the capabilities of stone tools used by our hominin forebears than we know about the metallic weapons of modern people. Swords, despite how we fetishize them in myth and history, are poorly understood when it comes to the damage they do. But, building off a few previous studies, anthropologist Jason Lewis set about categorizing the damage various bladed weapons can inflict on bone.

Using six bladed weapons—a katana, an Arabian-style scimitar, a wavy blade broadsword, a Samburu short sword, a machete, and a hunting knife—Lewis scientifically hacked at cow bones and then analyzed the damage. Swords and knives caused different kinds of trauma, with the size, weight, and swing of the weapon leaving telltale patterns behind. In general, sword marks were wide and deep, and caused a large amount of damage to the sides of the cut. Sharp as they look, they leave behind messy signatures that look like blunt-force trauma rather than clean cuts. Knife marks were shallower and had a distinctive V cross section, fitting blades that are usually thrust forward to stab. And working from research like this, the anthropologists looking after Richard set about investigating what befell him.

It's impossible to tell the order in which the wounds were inflicted. None of them overlap with each other, and at least two could have been deadly—that is, if Richard wasn't dead already. But what they clearly demonstrate is the unabashed cruelty Richard's body suffered. There are two wounds on his jaw—a tool mark on the jaw and a cut mark near the chin. Neither of these would have been fatal. Nor would have the injury caused by a 0.3-inch hole in Richard's right upper jaw that punched through the bone, but being stabbed in the face with a square-bodied blade surely wouldn't have been fun. Then, on the back of the skull, there are a pair of "shaving-type injuries" on the parietal bones. Whether these were made with the same weapon isn't clear, but this would have been a bloody wound, slicing the scalp and exposing the bone beneath. You can still see the striations from the blades on the surface.

It's hard not to cringe reading the next part of the report. There's a keyhole-shaped injury at the crown of Richard's skull, over the sagittal suture. "This injury was associated with an interior inner table injury: two bone flaps were pushed inwards towards the location of the meninges and brain," the anthropologists write. This was not a glancing strike, but "seems to have been caused by an oblique blow from a weapon delivered from above." Whatever weapon was being used punched through the top of the skull, damaging bone, the meninges, and maybe even the brain itself. But as we know from trepanation, such injuries don't have to cause death. At least not immediately.

Then, on the underside of the skull, there were two injuries that were somehow even worse. A yawning hole in the skull in the area of the right cerebellum—down and to the back—suggests a strike with a sword or a halberd. Then there's an injury in the area where the spinal cord exits the skull—the foramen magnum. The damage pattern suggests that "the tip of an edged weapon had penetrated through the bone and the brain and as far as the skull inner surface opposite the point of entry," about 4.1 inches. These were serious wounds. "If inflicted in life, either of the injuries on the inferior aspect of the cranium could result in subarachnoid haemorrhage, injury to the brain, or an air embolus." Technical terms for terrible events no one wants to experience. "The injuries are highly consistent with the body having been in a prone position or on its knees with the head pointing downwards." Richard had his head down, exposing the back of the skull and neck. These were strikes against a defeated king.

And that was just the skull. One of Richard's ribs showed a cut that came from behind by a fine-edged dagger. He may have already been dead by then, stripped of armor and totally vulnerable. Even worse was what the archaeologists called the "humiliation wound." There's a slice reaching across Richard's lower hip bone, separating part of it from the rest of the pelvis. The details of the cut show that it went from back to front. "Reconstruction of the pelvis with CT-generated images and the angle of the wound indicated that the weapon entered from behind, through the natural space created between the sacrum and the greater sciatic notch." The way the paper talks about this embarrassing end point is necessarily subdued. "In life, this injury could have caused damage to the internal pelvic organs including the bowel. This area is highly vascular, and, if inflicted in life, this wound could have caused substantial bleeding which could have been life-threatening." This is consistent with history; after death, Richard's body was slung over a horse and "suffering insults."

Exactly what happened on that field at Bosworth, no one can say. That knowledge died with those who were there. But the anthropologists were able to outline what likely happened to Richard. The king probably abandoned his horse, as legend holds, and lost the fight. He probably had his armor on when he died. The researchers note there were no defensive wounds to his arms or hands, as would be expected if he was still alive when stripped. What happened next is unknown, but the wounds to the back of his skull suggest that he either lost his helmet or had it removed by his captors, then was killed as he looked down. The point was to mostly leave his face untouched. He still needed to look like himself for public display for the Tudors to usher in their new order. Plus, it reduced the chances of someone looking more or less like Richard trying to claim the throne. Afterlife embarrassment was protocol.

Endings are clean on television. Cumberbatch's Richard III meets a quick end at the Battle of Bosworth, speared through the heart. We leave him on the battlefield, easily vanquished. Justice is done and Henry VII ushers in a time of peace now that England is free of Richard's madness. But the real bones of Richard III tell a very different story. Regardless of whether this king was good, evil, or generally as craven as any given member of medieval aristocracy, he met a painful, protracted end at the hands of his enemies. If *The Hollow Crown* had dared to reenact the scene, viewers would have found themselves sick. We are a cruel species, and the marks of our depravity are etched onto Richard's bones. Whether he was similarly cruel, his bones don't say. They can't tell us whether he was maniacal or merciful, murderous or misunderstood. Good and evil do not wind their way into our bones. In fact, the greatest exoneration—or, more likely, condemnation—of Richard would come by finding those long-lost nephews of his, the two princes who were said to have been murdered in the Tower of London. If their skeletons could be found, and their bodies scanned for signs of early and unwarranted death just as Richard's have, then perhaps the most heinous rumor surrounding Richard III could be resolved.

In the end, Richard III was given a second burial that was far more ostentatious than the first. After a procession through town, he was buried in Leicester Cathedral on March 26, 2015, attended by members of the British royal family and even Benedict Cumberbatch himself. Perhaps it's not what he envisioned, but it was a burial fit for a king. But such a case is truly extraordinary, and many excavated and exhumed remains are never returned to a place of rest. Instead they form collections in museums the world over, and many still speak to the ways the dead have been used to mistreat and oppress the living.

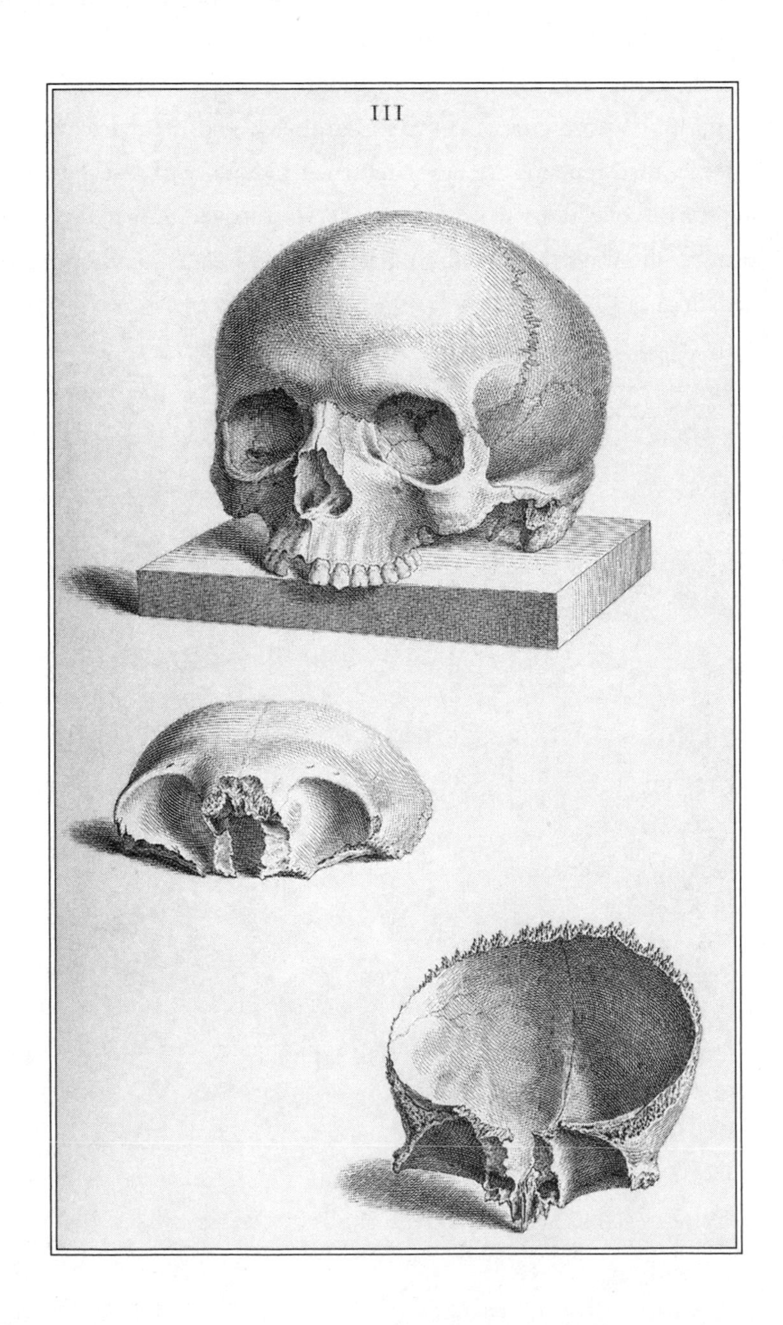
III

EIGHT

BONES OF CONTENTION

The skulls stared straight ahead, each cranium an impassive island of bones. These were people once. They still are. But the tidy setup in the museum hall didn't give that impression. They had become things, stripped of flesh as well as most of the details about who they were.

Each skull had been reduced to its ascribed identity. Some bore labels pasted onto their frontal bones. "Negro, born in Africa," "Chinese, Male," and "Peruvian of the Inca Race." Others displayed jottings across their crania made at different times. Human, but denuded of their humanity. Instead each became an anatomical totem for the racist foundation of early American anthropology. These are the skulls of Samuel Morton.

The collection, held at Philadelphia's Penn Museum, is huge. Morton himself had assembled 867 skulls by the time of his death in 1851. After that, his colleague James Aitken Meigs continued adding until the total rose to more than 1,225 crania. For a time, this macabre collection was open to the public. When the collection was housed at the Academy of Natural Sciences, across the winding Schuylkill River, visitors could peer at the skulls for free on Tuesdays and Saturdays. What's now on display is a peek at the hundreds upon hundreds of additional crania held behind the scenes, a small selection of skulls lined up in sleek, well-lit display cases with accompanying signage.

It's strange to stand before the skulls. The display isn't much different from that of the Mütter across town or classic collections like the Hunterian in London. It's both intimate and cold, the sterile nature of the exhibits in hard contrast with the fact that the skull is the most human part of our skeleton. The osteological faces still carry personality, even so long after death, and those capacious domes behind the faces held the brains that made each and every one of those people who they were. Even thinking of them as people who lived and died long before I was born changes the feel of the exhibit space. Morton had acquired every skull he could because of what those remains would say about people—about us—and yet the crania are displayed just as a fossil or ancient pot would be. Seeing Morton's tools next to the skulls—and, in one case, bracing a cranium for measurement—adds another layer of scientific detachment. Each of these unique people was reduced to data points that, in Morton's nineteenth-century mind, represented their race. They would not be judged on the lives they led or their actions, but merely on how much lead shot their braincases could carry.

This is where our story takes a darker turn. We've traveled through the origin of our skeletons, from the basic anatomical layout through the origin of bone and how it reacts to the world through the way pathologies speak to the lives we lead. But now we're deep into the afterlife of bone, where the living project their ideas about life onto the perished. The picture is often unpleasant, not because of what's done to the bodies themselves, but because of the misguided ends bones can be put to. We're still left with the consequences of when an obsession with race fueled the beginnings of American anthropology, and Morton's skulls provide us with a case study for how we often project our thoughts and ideals onto naked bone.

Not that Morton was the first to do so. To understand what the physician was trying to do with his skull collection, and the harm enabled by an exercise labeled as objective science, we need to rewind to the turn of the nineteenth century, when anthropology and related sciences were beginning to coalesce from their amateur beginnings. Physiognomy, phrenology, and craniometry all attempted to be the objective sciences of their time, only to precipitate terrible consequences.

Sciences don't have single beginnings. They don't appear fully formed out of nowhere. The sciences we've invented are always set against the background of what was previously thought and the time they originated, our attempt to objectively understand nature always framed by human culture. This means that any scientific story has multiple possible starting points, and many sciences began with ideas that have since been tossed into the bin of pseudoscience. In this case it's fitting to go back to Johann Caspar Lavater.

A Swiss polymath working in the late eighteenth century, Lavater made a name for himself in everything from poetry and philosophy to theology. In the realm of science, he's primarily remembered for his work on physiognomy—trying to judge character by appearance, the anatomical equivalent of judging a book by its cover. If you could properly read someone's body shape, the argument went, then you could understand their personality, all based on the supposition that nature cannot lie and is easily observed as essential truth. Lavater's interest in finding signs of God in the flesh drew him to the idea, for surely the Almighty delicately crafted all life and made the external a guidepost to the character inside. "Not even the skin of a flea was made by chance," he declared. This gave the experienced and educated physiognomist the power to instantly judge the character of a person at a glance, with the details of the mouth, chin, cheeks, and even hair all acting as indications of behavior. The idea took off. Lavater's lavishly illustrated books were a wild success, running through multiple editions in German, French, and English. A popular handbook version was even created so that readers could readily judge others at a glance. It's no surprise that this guide aligned with what Western European males found attractive or as signifying virtue. Lavater proposed that people with high foreheads were more likely to understand you, for example, while those with lower domes were much more likely to fight with you.

What Lavater was trying to construct was a typology, a standardized set of categories for the human body that could easily be read and would always ring true. Phrenology picked up where Lavater's physiognomy left off. Founded by the German physiologist Franz Josef Gall in the 1790s, phrenology asserted itself as a science that would dispel the mysteries of the human mind and behavior. These self-styled experts, including Gall, were confident that the brain is the wellspring of the mind, and that the mind is made up of complementary but discrete faculties. Each of those faculties, in turn, is tied to a specific organ of the brain, and the size of the organ determines how much power or influence that organ has. (These organs, called centers, were not parts of the brain like the frontal lobe or hippocampus; they were tied to their function and were believed to determine everything from animal urges such as destructiveness to the highly valued center of virtue.) The shape of the brain, therefore, is determined by the arrangement and form of these various organs, and the human skull has a tight fit to the brain. It follows, then, that the shape of the skull records the sizes and shapes of the different brain organs and can be used to quickly and easily profile someone's mind just from its appearance. Phrenologists were taking God's exhortation from Job to "speak to the earth, and it will teach you," adamant that what nature had manifested provided all the critical information about our inner lives. The mind was no longer a matter for philosophers or even theologians. Gall and his successors were establishing a new typological system based on observation and measurement. "The pronouncements of the phrenologist were said to be true," historian John van Wyhe writes, "because they were scientific facts drawn from an unerring, constant Nature."

Through books, debates, and lectures, Gall made a prominent name for himself in the upper crust of Western European society. It didn't matter that other scientists told him he was wrong and that theologians were often aghast at the idea of putting science where they believed an inscrutable god should be. After his death, colleagues such as Johann Spurzheim continued Gall's legacy and hyped phrenology even further. Gall always insisted that the practice was special knowledge that could not and should not be possessed by the public, but its popularization in the nineteenth century allowed phrenology to mutate into a sideshow attraction. Phrenologists enjoyed much the same status as nineteenth-century spiritualists, mesmerists, and other purveyors of cultural fads, the equivalent of today's phone psychics and palm readers, but using the shape of people's heads.

Phrenology claimed that people's minds could be read at little more than a glance, and its proponents insisted that this would lead to the end of all sorts of social ills. Phrenologists with a progressive bent, for example, clamored for an end to capital punishment. The cause of murder and other heinous crimes was not sin or some inscrutable force. Murderers and criminals obviously had brains that had developed in the wrong way, the organs of their gray matter causing terrible behaviors that led them to crime. Biology had determined their fate, and who could punish them for that? In more practical matters, phrenologists insisted that knowing your own mind—and those of people around you—would lead to a richer, better directed life. A pamphlet by phrenologist H. Lundie advised:

> If about to marry, Phrenology will point out to us the characters, dispositions, and tempers of those with whom we are about to be united for life. Have we sons and daughters—and do we wish them to learn those trades, callings, and progressions best suited to their various organisations? If so, Phrenology will point out to us their various capabilities, and prevent that cruel disappointment which is to be met with in every day's occurrences . . .

But, as van Whye observed, much of this posturing didn't ring true. Progressive phrenological devotees could use head bumps to argue for social reform, but phrenology itself was value-free and could be used to a variety of ends. Gall and Spurzheim were primarily interested in using their special knowledge to maintain their careers, and the idea that the practice was largely a social movement comes from the books, pamphlets, and letters of those who had a vested interest in maintaining its popularity. In the end, phrenology was about power. This accessible "science" gave adherents a quick way to read others and determine their destinies; the possible paths of one's life could be opened or closed depending on the arrangement of organs one was born with. And this appealed to the socially conservative as much as to the progressive. Calvinists, for example, jibed with phrenology because it was a biological expression of predestination. If God created organs of greed and desire in the brain, and these organs were enlarged in some people, then these poor souls were doomed from birth and could never truly change their ways. The virtuous were simply born that way.

That's hardly all. As phrenology gained popularity in nineteenth-century England, it readily spread to British colonies around the world. In these places, it was not a discipline of social reform. Instead it was an empirical tool for maintaining a stranglehold on society. In Australia, historian Russell McGregor notes, "Phrenology played a significant part in fostering the notions that Aboriginal mental powers were limited and their prospects for improvement slight." George Combe, the UK's most prominent phrenologist, looked at the collected skulls of Aborigines and concluded that these people were "distinguished by great deficiencies in the moral and intellectual organs." This was biological determinism at its worst, entire cultures written off and further oppressed because science said they would never assimilate into European society.

Phrenology was used to much the same ends in colonial South Africa, where social conservatives embraced it as a scientific basis for continued discrimination, violence, and suppression. Between 1779 and 1879—the Cape Frontier Wars, also called Africa's 100 Years War—European colonists constantly clashed with South Africa's indigenous Xhosa tribes. European soldiers made a habit of taking heads from their victims—a practice that the Xhosa learned from the Westerners and eventually took on themselves—and these grisly trophies were then used as evidence that the Xhosa were inferior and deserved their place in society.

The oppression suffered by these people did not originate with phrenology, but Gall's and Spurzheim's popular categorization system added new, seemingly empirical evidence for those who wanted to maintain their power. Politicians and the populace didn't have to argue on the basis of religion, history, or philosophy. They could point to bumps on stolen skulls and assert that the disenfranchised were taking up the only role in society they would ever have. Instead of unlocking the secrets of the mind, phrenology was used as a means to maintain control. It was against this background that Samuel Morton began his massive skull collection, and his supposedly objective categorization of people would be used to much the same horrific ends.

While Morton didn't describe himself as a phrenologist, his anthropological interest nevertheless shared a focus on human skulls and their measurements. His goal was to develop a standardized way of distinguishing groups of people according to uniform and observable measurements, symptomatic of a struggle many sciences were having during the time: physics envy.

Scientists and naturalists of the nineteenth century were in awe of physics. It was amazing just how much of the universe and its motions could be boiled down to seemingly immutable laws. And because physics was early out of the gate during the nineteenth-century science boom, its practitioners were able to convince other scientists that physics was a hard science and the standard other sorts of investigations would have to reach if they were to be taken seriously. Not that this was entirely bad. Measurements and data make scientific proposals testable and repeatable. That ability to constantly check and question is the very heart of science. Otherwise conclusions don't differ from dogma. And it was far better to have archaeologists and anthropologists taking detailed notes and measurements than to have them tearing through burial mounds and rifling around for whatever caught their eye, as had been the standard up until then. The pressure from sciences such as physics required that anthropology be carried out in a measured and systematic way. Samuel Morton certainly tried, using his position as a physician at Pennsylvania Medical College to accumulate as many skulls as he could.

The inspiration for Morton's grim collection came to him in 1830. Drawing from the racial ranking of German anatomist Johann Blumenbach, Morton set about putting together a lecture titled "The Different Forms of the Skull as Exhibited in the Five Races of Men." These, as outlined by his European colleague, were Caucasian, American, Malay, Mongolian, and Ethiopian. As you might suspect, these were arbitrary labels that tell us more about the Western scientists who invoked them than they do about any actual sets of characteristics of real populations. Morton was confident that these divisions were correct, but he didn't have an adequate skull collection to demonstrate the point. So, leveraging his prominent position at the college, he started writing to colleagues and contacts to build up his very own skull library. It worked. Within three years Morton had acquired nearly a hundred skulls, and the osteological trickle was still going. The collection became so famous that it was almost a point of honor to contribute to it, even if that required the terrible practice of robbing graves to obtain skulls for the Philadelphia doctor.

The famous collection formed the basis for the books Morton is most remembered for: *Crania Americana; or, A Comparative View of the Skulls of Various Aboriginal Nations of North and South America: To Which Is Prefixed an Essay on the Varieties of the Human Species* (1839), *Crania Aegyptiaca, or, Observations on Egyptian Ethnography, Derived from Anatomy, History, and the Monuments* (1844), and his final summation, *Catalogue of Skulls of Man and the Inferior Animals* (1849). All three sound appropriately lengthy and stuffy for nineteenth-century scientific treatises, but these old books became hits among scientists and politicians alike. That's because Morton, intentionally or not, gave those in power a scientific basis for justifying their cruelty to other humans.

Morton's main anthropological interest was to measure skull capacity in each supposed race and compare them with each other. Exactly why, no one really knows. Part of his reputation as the objectivist of his day was that he didn't do much to analyze or interpret the data he collected. The differences between the five assigned races seemed to be self-evident, and so Morton merely focused on making his measurements standardized and repeatable, presenting them as biological fact. (Phrenologists did much the same, not bothering with analysis or theory but taking their system of the mind as a given.) And the way Morton did so would eventually underscore the bias behind his research.

In order to get his skull measurements, Morton initially packed the vacant crania to capacity with white mustard seed and then poured the seeds back out into a graduated cylinder to get a read of the skull's internal volume. It was a smart method, especially using a fine-grained filler to occupy as much of the space as possible. But Morton soon realized that this method had an unavoidable flaw: multiple measurements on the same skull with the mustard seed weren't yielding consistent results. Despite sieving them to try to obtain seeds of the same size, they still varied; measurements from the same skull could be off by several cubic inches. The seeds simply couldn't offer Morton the accuracy he needed. So he switched to BB-size lead shot instead. Despite being larger, they worked much better and his measurements became more consistent.

Morton's cranial capacity results came out just as anatomists of his time would expect. Now, remember that none of these racial categories have any real biological meaning, and which skull Morton attributed to what group was entirely subjective and somewhat dependent on information he received about where the skull came from (which, given that many were obtained by grave robbing and similar means, might not have been accurate). Applying these deeply flawed methods, he reported that his "Caucasians" came out on top, his "Mongolians," "Malays," and "Americans" in the middle, and his

"Ethiopians" on the bottom. (Morton's second book, on the skulls of Egypt, showed a similar trend, with races he labeled as "Caucasian" at the top and "Ethiopians" on the bottom.) This was presented as objective fact. He took measurements, and here they were. But the implications of his research were clear. If Morton wouldn't openly theorize, others were more than happy to do it for him.

Science is not a matter of simply discovering facts and putting them on a shelf. There is a natural world that can be understood, but *how* those facts are understood requires theory. Facts, in other words, are always analyzed in the context of what we think we know or expect about how the world works. To borrow from a field closer to my own, Charles Darwin once lamented that his fellow naturalists chided him for spending so much time on theory. If he wanted to convince people of his outlandish evolutionary platform, he should speak in terms of plain facts only. But Darwin knew this was totally absurd. "About thirty years ago there was much talk that geologists ought only to observe and not theorize," he lamented in a letter to his friend Henry Fawcett, "and I well remember someone saying that at this rate a man might as well go into a gravel-pit and count the pebbles and describe the colours. How odd it is that anyone should not see all observation must be for or against some view if it is to be of any service!" If a fact is going to help us understand nature at all, then, it has to be part of an argument. And in the first half of the nineteenth century, the scientific argument regarding humanity was centered on race.

The standard view at the time, even among naturalists, was that God created all of life. The question was whether there was one creation or many, and what purposes those creations served. Hence, in the appendix to Morton's *Crania Americana*, the famous Scottish phrenologist and Morton fan George Combe wrote that what the Philadelphia physician had discovered could not be looked at simply as a mass of new facts that stood alone. Those cranial measurements carried weight about the intellectual and moral abilities of the people the skulls once belonged to. "If this doctrine be unfounded, these skulls are mere facts in Natural History," Combe wrote, "presenting no particular information as to the mental qualities of the people." No, Combe argued, this information had deep implications for the entire history of humanity, with no shortage of evangelists ready to take Morton's measurements to mean there was now scientific proof that whites were superior. On top of that, the American brand of physical anthropology in those days before evolutionary theory was particularly enamored with the idea that there could have been separate creations, with different races created in different places. By this logic, some of the more extreme advocates for racial separation suggested that "Caucasians" and "Ethiopians" were effectively different species. History as seen through a white lens was brought to bear, too. Given that even the ancient documents of Egypt showed dark-skinned people as slaves, and such records were effectively as old as white people were willing to recognize of history (waving aside the histories of aboriginal peoples), they concluded that black people had always been slaves and this was their role in the natural order. While Morton himself didn't make such arguments in his books, he did nothing to stop his friends and skull suppliers from using his book as the definitive document bolstering white superiority.

It's difficult to look back on the histories of Morton and his colleagues and not feel rage start to boil over. What they were so sure was biological fact was nothing more than racist fantasy. Morton has generally been held apart from the various naturalists and politicians who used his works as a justification for slavery and the social order whites violently inflicted on the country. He only collected data and shared what he found. It was up to others to analyze, provide commentary, and try to defend the evils of slavery. But as historian William Stanton pointed out in his book on this dark time in American history, *The Leopard's Spots*, as Morton continued his writings, he left no doubt as to his views on the rest of humanity. In a letter he compared "the noblest Caucasian form" to "the most degraded Australian and Hottentot," making it clear how he personally saw the world. He just couldn't accept that humanity had a common origin, or that the circumstances of life alone could modify the human skull.

While he was the scientific star of his time, however, Morton's work was soon forgotten. The idea of evolution by means of natural selection—simultaneously thought up by Charles Darwin and Alfred Russel Wallace, articulated by Darwin in *On the Origin of Species* in 1859—made suggestions about different centers of creation and the separate invention of species obsolete. The Civil War and Emancipation Proclamation further undercut Morton and the more vocal purveyors of the idea that skull capacity was destiny. Racism and inequality didn't end, not at all, but the particular science-based brand of race ranking that Morton and his contemporaries were focused on faded as society shifted in the wake of America's bloodiest conflict. And so Morton's work became a footnote in the history of science—that is, until the late twentieth century.

Paleontologist and essayist Stephen Jay Gould was the one to kick the hornet's nest, arguing in his 1981 book, *The Mismeasure of Man*, that Morton had an unconscious bias against black people that seeped into the Philadelphia physician's supposedly objective measurements. The key clue was the difference between the mustard seed and lead shot values. When Morton switched from seed to shot, Gould pointed out, all of his volumetric measurements shot up. But his measurements for his "Ethiopian" category rose by a higher amount than any others—that is, he might have been unintentionally fudging the seed measurements because it was possible to pack more or less seed into skulls he had already divided by race, but lead shot wouldn't allow for the same bias to creep through.

The keepers of Morton's skull collection didn't take kindly to Gould's characterization. In 2011, a team of anthropologists led by Jason Lewis printed a late-arriving reply to Gould's charge—the new study claiming that Morton had taken his lead shot measurements accurately, something Gould never disputed—which further catalyzed additional responses. Papers zinged back and forth over whose numbers were better—Morton's or Gould's—but, as historian Jonathan Kaplan and colleagues pointed out, it doesn't really matter whether Morton accurately took his measurements. His whole program was flawed from the outset, working within a racial system with no biological meaning and with skulls collected from dubious sources. We don't even know why Morton wanted to measure average cranial capacity or what he thought this would prove. He took accurate measurements with lead shot. Fine. But collecting data is just the start of science, not the end, and Morton clearly had no qualms about his data set being used to reinforce the racist notions of his time. "It is hard to see how a set of skulls, collected unsystematically, often of uncertain provenance . . . and identified in premodern fashion with no way to tie them back to meaningful biological groups," Kaplan and colleagues observed, "could be usefully deployed to answer any meaningful question about the larger populations from which they were drawn."

Anthropology today isn't like it was in Morton's time, however, and major anthropological associations repudiate the notion of biological race. So how did we get from early practitioners like Morton to where we are now? Anthropology itself would end up showing how baseless such racist views are, and the living would be the basis for undoing what had been argued from the dead.

The change came in the early 1900s. Like most physical anthropologists of his day, Franz Boas was curious about the origins of racial types. How many were there, and how did they become established? Working at the beginning of the twentieth century, he was in the perfect position to investigate those questions. Boas was conducting research at the American Museum of Natural History then, just when New York City was seeing a massive influx of immigrants from abroad. The local population swelling the city's neighborhoods were Boas's test subjects. If races were distinct and fixed, he reasoned, then the American children of immigrants should show the same characteristics as their parents. But if those children differed significantly from their parents, who grew up in other parts of the world, then it meant that "races" were actually plastic and were not divisions fixed by heredity.

So Boas started measuring. In a 1908 study, he took cranial measurements of Russian Jewish boys enrolled in New York's public schools. He didn't find what he was expecting. He thought the boys were going to fall within the range of measurements he had collected from Europe. Instead, even in first-generation offspring, Boas detected significant differences in cranial details, and so the traditional view of fixed, immutable races could not be upheld; the categories didn't exist. But Boas wanted to be sure. (He was concerned, for example, that his choice of subjects in public school might have influenced his results in that these children had more advantages than those of poorer classes, with better nutrition and less disease affecting their anatomy.) To that end, he and his assistants took about twelve hundred sets of cranial measurements per week from newly arrived European Jews, Bohemians, Sicilians, Poles, Hungarians, and Scots and their children, totaling a data set of more than eighteen thousand profiles. The results were unmistakable. The skulls of American children differed significantly from those of their parents, showing that environmental influences—such as nutrition, disease, and stress—could affect one's cranial shape. These factors caused a great deal of variation and defied the simple racial categorization system that had been popular for so long. This was an idea eighteenth- and early nineteenth-century naturalists had suggested, but it had been steamrolled by the insistence of the first generation of American anthropologists that races were created separately and remained discrete regardless of environment. As Boas himself wrote in 1911:

> Head form which has always been considered the most stable and permanent characteristics of human races, undergoes far-reaching changes coincident with the transfer of the people from European to American soil. . . . These results are so definite that while heretofore we had the right to assume that human types are stable, all evidence now is in favour of a great plasticity of human types, and the permanence of types in new surroundings appears rather as the exception than as the rule.

But Boas felt he was alone in his view. His colleagues still maintained the primacy of race, and in the public realm, Boas's findings were misconstrued as evidence that a new American race was beginning to form and join the others. Anthropologists held tightly to the mythology of race and the ranking system that came with it. Americans of European descent characterized themselves as being of white, noble Nordic heritage that had to defend itself against dilution from the varied immigrants arriving onshore. It was only after World War II, when the ultimate evils of racist views were made clear by the Holocaust, that anthropologists began to forcefully back away from the racial divisions their science had been founded upon. In 1950, as the UNESCO director-general, biologist Julian Huxley convened anthropologists and sociologists to draft a statement on the science of race. While the experts still outlined what they saw as divisions in humanity, the main point of the statement was that "For all practical social purposes 'race' is not so much a biological phenomenon as a social myth," with no difference in the innate abilities of any "race" compared with any other. The statement wasn't universally accepted—anthropological traditionalists and even some former Nazis were so vociferous in their defense of race in scientific journals that UNESCO assembled a second group of biologists to draft a follow-up statement—but it still marked a turning point for anthropology. At least some twentieth-century researchers were trying to push back against the discipline's traditional fixation with naming, categorizing, and ranking races.

This was difficult work, especially because, as biological anthropologist Jonathan Marks points out, "physical anthropology is burdened by the weight of phrenology, polygenism, racial formalism, eugenics, and sociobiology," the foundation of the endeavor being the search for biologically determined differences. Anthropology didn't invent racism, but it certainly fueled it through the nineteenth and early twentieth centuries. This could only change slowly and painfully, as the horrors of World War II and then the burgeoning civil rights movement in postwar America made abundantly clear the injustices and imbalances that a fixation on separating races had helped foster. The development of genetic studies has only underscored that anthropology's early racist conclusions were subjective and relied largely on the perspective of whoever was interpreting humanity. Studies of our species's DNA, for example, have highlighted that there is more genetic diversity *within* human populations than between them. On every single point, from our DNA to our very bones, there is no biological trace of race. Try to make any division you want based on skin color, height, eye shape, or any other feature and you'll find people who don't fit into whatever typology you choose. Just as phrenology was, the obsession with collecting and measuring skulls within a racist framework was based on the concept of those with power maintaining their grip. This has been apparent from the outset, or, as W. Montague Cobb, a Howard University anthropologist and the president of the NAACP from 1976 to 1982, wrote:

> The three factors of commercial interest, ignorance, and pride of conquest thus combined to create in the mind of the European civilization of the day an impression of biological inferiority as regarded the black man. There is little occasion for surprise that early physical anthropologists seemed to accept the concept of the stratification of human races, with the white race at the top, as biologically sound. Nor is it remarkable that there should have been instances when able men adduced anatomical evidence in support of this view, either because of sincere conviction, or, unconsciously, to furnish justification for a trade which currently represented powerful economic interests.

Bones hadn't been used to find the truth of humanity, as Morton said he aimed to do. They were used to suppress and subjugate, and the echoes of those efforts affect us even now.

There are some who simply shrug and say that anthropology has moved on, that this is merely how science proceeds. Anthropologist Kenneth Kennedy wrote, "If the apparent 'racism' of physical anthropology of the pre-1950s seems to characterize a very different discipline from the one we practice today, this should not evoke embarrassment. Western astronomy has its origins in astrology; early chemistry was alchemy; medicine emerged from shamanism; and biology had its early home in natural theology and the concept of the chain-of-being." But that's rather cold comfort to the millions who suffered and died—and are still discriminated against today—in part because nineteenth-century anthropologists enthusiastically propagated the myth that races are real and define our place in the world. Those efforts forced painful, still-stinging divisions that we are coping with right now. Supposedly science-based racism has been popular in conservative politics and is even experiencing a resurgence. Insults and arguments from white supremacists sound a lot like what advocates for slavery were saying 150 years ago. It's difficult to escape this ugliness. In early 2018, for example, a new genetic analysis of a ten-thousand-year-old skeleton from Britain known as Cheddar Man indicated that this person had dark skin. This wasn't a huge surprise—humanity evolved in Africa before spreading through the rest of the world—but, in a modern Western political climate suffering from a resurgence of racism, xenophobia, and political isolationism, angry commenters crawled out of their corners of the internet to fume over the fact that one of the earliest people to call ancient England home had a dark complexion.

The most unsettling aspect of our bones isn't life sliding into death. It's what the living do to the dead, using the bones of the deceased to consolidate and maintain systems of power. Science may have moved on, but the scars remain. And rather than hide from this, anthropology has a duty to honestly assess its own past and the impact of what's been done in the name of science, and why many people have good historical reason to not immediately trust that science has their best interests at heart. There is no instantaneous fix for this, and building trust from such a terrible history is no small challenge. This leads us to an uncomfortable question that lurks in every anthropological collection and archaeological study of human remains: Who are the rightful keepers of the dead?

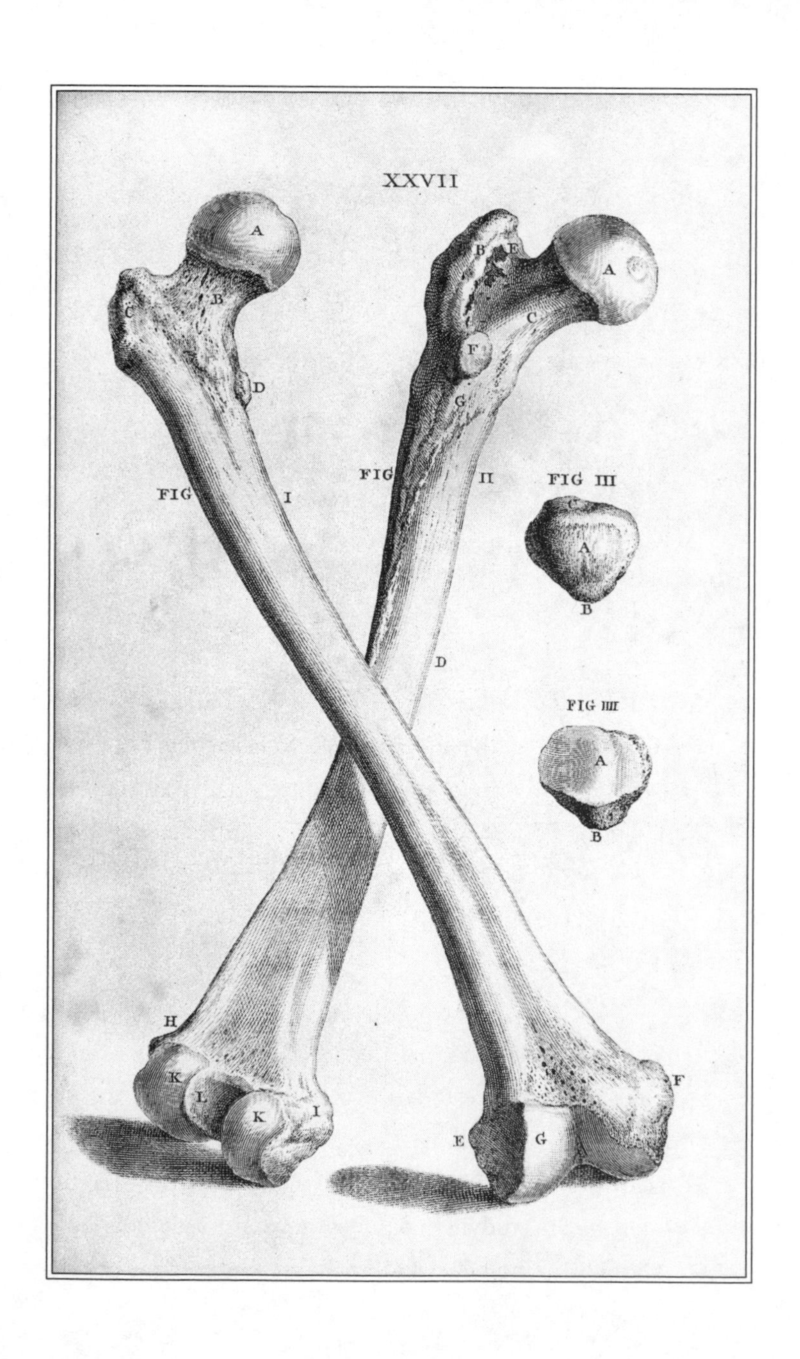
XXVII
FIG I
FIG II
FIG III
FIG IIII
A
B
C
D
E
F
G
H
I
K
L
K
I
D
E
G
F
A
B
C
A
B

NINE

SKELETONS IN THE CLOSET

Early on the chilly morning of February 18, 2017, at an undisclosed location on the Columbia Plateau in the northwestern United States, more than two hundred people gathered to return a skeleton to the earth. This gathering was more than a burial; it was a victory. As Chuck Sams, communications director for the Confederated Tribes of the Umatilla Indian Reservation, said, "A wrong had finally been righted."

For two decades, the Yakama, Wanapum, Umatilla, Colville, and Nez Perce nations had argued that a disinterred skeleton was all that was left of one of their ancestors. But they had been just one party among many declaring a right to the bones. It seemed almost everyone had tried to claim this skeleton for themselves, from anthropologists interested in studying the nine-thousand-year-old remains to white nationalists who believed it proved the Americas belonged to a different people before the First Nations. No matter what side of the fight anyone was on, everyone felt bruised by the outcome. The prolonged battle was a painful lesson in the importance of who holds power over interpretation of the past, and who—if anyone—owns the dead.

In America, there is perhaps no group this question has affected as strongly as it does Native Americans. Science, in their history, is not just another way of knowing about the world, but an institutionalized force that often saw their culture and their bodies as objects to be collected. The history of science has normalized the acquisition and study of ancient skeletons, particularly from indigenous peoples. When the aggrieved have protested, scientists and authorities have often argued that the burden of proof is on them to show the accuracy of their claim while the bones they want returned are held out of reach. Scientific tradition holds a power it doesn't always want to recognize.

The skeleton scientists would come to call Kennewick Man, and whom the coalition of nations known as the Confederated Tribes of the Colville Reservation called the Ancient One (as will I), had been interred in an intentionally dug grave for more than nine thousand years, laid to rest by those who knew him best. Everything changed on July 28, 1996, when a skull turned up along the Columbia River in Washington. The police called in anthropologist James Chatters, and when he and others working at the site found more bones, Chatters rushed to get an emergency Archaeological Resources Protection Act permit from the Army Corps of Engineers, who oversaw the land the bones were found on. Eventually the team working on the site would turn up most of a human skeleton. Skull, spine, ribs, hips, arms, hands, legs, and feet were all there; only a handful of the standard 206 bones that make up the human skeleton were missing.

Naturally, the immediate question on everyone's mind was who this person was. There were no clothes or objects to give a clue to when in history he had died and been buried. Chatters sent one of the bones of the left hand to the radiocarbon laboratory at the University of California, Riverside, which estimated that the bones were between 8,340 and 9,200 years old. This left no doubt that the skeleton predated European contact with Native Americans, making it one of the oldest and most complete ever uncovered in North America. This was a person who walked the Pacific coast not so long after mammoths and saber-toothed cats ceased to roam.

But that initial discovery would challenge a newly enacted provision meant to protect such remains. The age of the bones excavated by Chatters meant they fell under the provisions of NAGPRA—the Native American Graves Protection and Repatriation Act—and according to that law were automatically assumed to be from America's aboriginal peoples. Upon learning the radiocarbon date, the Army Corps of Engineers seized the bones, put them in an evidence locker at the local sheriff's office for safekeeping, and ordered a halt to any further testing of the remains, marking the beginning of a painful conflict that would stretch over the next twenty years.

NAGPRA is the kind of provision that seems to make no one totally happy. At its inception, scientists complained that they were being denied crucial information about America's ancient people. Some of those people whom NAGPRA is supposed to protect, on the other hand, felt that the law was too much on the side of researchers who seek to hold on to the remains of their kin. Almost thirty years on, it seems that the law has generally been a positive for researchers as well as Native Americans—emphasizing communication, collaboration, and transparency—but the reason such a provision was necessary at all rests on a long history of cultural theft. The regulation, enacted in 1990, cannot undo the decades of grave robbing and appropriation that built many a museum collection. At most, it can send some of what was stolen back and provide a set of rules to prevent further damage.

As initially envisioned, NAGPRA was supposed to protect the remains and cultural items of Native Americans found on federal or tribal lands after November 16, 1990. These pieces of the past are to be returned to the lineal descendants, but the older the remains are, the more difficult it is to prove a connection. The tribes who claim the remains have to provide proof of their relationship, which can sometimes lead to impasses. Some ancient remains may belong to bands or tribes driven to extinction, or to existing groups who aren't recognized by the federal government. In some cases, a group might object to the kind of DNA testing or scientific study that might be needed to establish kinship according to the law. This is where politics and history run headlong into tensions between science and culture, where empirical knowledge and cultural belief are weighed against each other and science is often given more heft. Even if it's true that scientific research is necessary to make connections between ancient remains and people living today—at least as far as the law is concerned—this can feel like an insult to people who assert a relation based upon the traditions and stories of their own culture. These matters are extremely delicate, carrying the weight of centuries of European bigotry and oppression against Native nations. The Kennewick case blew everything up by highlighting a weakness in the new regulation.

NAGPRA wasn't meant to handle an ancient skeleton that lacked any obvious connection to today's Native people, yet the Ancient One became a symbol of the new legislation's deficiencies. NAGPRA's provision stated that any sufficiently ancient skeleton should automatically be categorized as Native American. There was—and remains—no hard proof that any other group of people or cultures were present prior to that time. But that's not what Chatters saw when he looked at the bones that had been uncovered.

Chatters thought the skull bones held hints of square-jawed European ancestry, or possibly even Asian connections. In fact, at first sight, he thought the skeleton belonged to a white settler in the area. When he learned the skeleton's true age, he still described the Ancient One as "Caucasoid" on the basis of a long face, square jaw, and narrow cranium. Then he set about making a facial restoration, which he presented at the sixty-third meeting of the Society for American Archaeology. The face Chatters showed off to his colleagues looked awfully European. Though Chatters was adamant this wasn't the message he was trying to send—"No one is talking white here," he told the anthropologists at the meeting—the new restoration was the spitting image of Patrick Stewart, ready to command as Jean-Luc Picard on the bridge of *Star Trek*'s USS *Enterprise*.

Facial reconstructions incorporate science. There are defined muscle attachments, details of fat and fascia depth, and other considerations that go into these re-creations. But they're still just hypotheses, subjective guesses based on data and measurements. It's impossible to definitively define a race or nation of descent from bones alone. There are no osteological markers of skin color or country of origin. An anthropologist can look at a skull and make a hypothesis based on a suite of features, but there's no foolproof way to make those kinds of attributions. And here was the Ancient One, brought back to life as a white man. He was therefore cast as a traveler who had come from somewhere else, someone with no connection to the Confederated Tribes of the Colville Reservation, who disputed the archaeologist's statements.

While there was no scientific grounding for the restoration Chatters presented, some took it seriously. The anthropologists who wanted to study the bones were convinced the Ancient One wasn't Native American—they claimed the skull's characteristics made it a mysterious outlier of unknown origin and that the public had a right to learn more about his past and what it might teach us about the history of the continent. The racially motivated Asatru Folk Assembly used the news of the Ancient One's supposed whiteness to say that this person had close European kinship. White nationalists had no qualms about jumping in, replaying arguments that had been buried a century before.

During the nineteenth century, as anthropology and archaeology were just starting to think about how to become sciences, there was a lot of hubbub about strange hills and earthworks found throughout the Midwest. Digging into them—roughshod at first, and later more systematically—showed that they were burial mounds filled with artifacts and bones. But who did they belong to? The obvious answer was that these burial mounds were made by Native

Americans hundreds or even thousands of years prior. But some experts couldn't believe that Native Americans had the sophistication to make such structures or that their cultures could have changed. People were supposed to be fixed in time, the entire evidence of human history to the contrary. Instead, some early archaeologists proposed that there were other people here before the Native Americans—an unknown, even more refined Mound Builder culture that had ties to Mexicans, Danes, Hindus, or just about anybody except America's indigenous people. The pernicious corollary to this archaeological argument was that Native Americans were not, in fact, native at all, and that the continent's original inhabitants were European kin. In this light, white expansion to the west and the efforts to drive Native peoples to extinction were recast as European descendants reclaiming what was once theirs. This played into the racist beliefs of the time that Native Americans were primitive and backward, and had chosen a crude life by rejecting the supposed improving influences of white civilization.

Of course Native Americans created the mounds. By the close of the century, the tools and other artifacts within had made that clear to the archaeologists. But it's difficult to look at those long-discarded ideas and not see shades of what transpired when that controversial skeleton was unearthed in Washington. Only now there was a tangle of expert interest and government bureaucracy against the backdrop of the history of discrimination science was trying to move beyond. The Army Corps of Engineers disagreed with the claims of Chatters and a group of eight additional anthropologists who wanted to study this "traveler" from unknown lands. The anatomy of the bones fell within the range of variation expected for

Native Americans, government scientists concluded, and so there was no reason to think that the Ancient One was some mysterious nomad from another culture. Thus, the Army Corps of Engineers started making arrangements to return the skeleton to the Umatilla tribe, who claimed kinship. Alarmed that science might lose a critical piece of our continent's past, especially if it added new detail to who lived in North America and when they got here, the outside anthropologists who wanted access to the bones sued. The case, *Bonnichsen v. United States*, dragged on for the better part of a decade.

The decisions made about the Ancient One hinged on a very basic question: Was this person Native American or not? If he was, then the law was absolutely clear that repatriation was warranted upon request. If not, then the law had no prescription for what to do. But the question itself is not clear-cut. "The conflict between Indians and anthropologists in the last two decades has been, at its core, a dead struggle over the control of definition," Standing Rock Sioux author Vine Deloria Jr. wrote. "Who is to define what an Indian *really* is?"

NAGPRA didn't specify how the Native American status of remains is determined, and the anatomy of bones themselves isn't that much help. Even so, there are anthropologists—particularly in forensic fields—who insist that races can be assigned to unidentified remains. In a paper published a few years before the Ancient One was found, anthropologist Norman Sauer asked, "If races don't exist, why are forensic anthropologists so good at identifying them?" Proposing a race for a deceased person's bones is a regular part of coming up with a forensic profile, Sauer wrote, and these were used in a trifold black, white, and Asiatic division to "reflect the everyday usage of the society" forensic anthropologists serve. This doesn't have anything to do with the question of biological race, he added, but is an educated guess that might help match denuded remains to a missing-person report. All the same, he concluded, "That forensic anthropologists place our field's stamp of approval on the traditional and unscientific concept of race each time we make such a judgement [on identification of remains] is a problem for which I see no easy solution." Perhaps some other term, like *ancestry* or *affiliation*, could be used to denote features that correspond with particular groups of people as determined by place of origin and culture. But as archaeologist David Hurst Thomas wrote, even the concept that there are certain features that can be used to make such distinctions is problematic. Even a meticulously documented collection of skulls divided into racial categories would still be based on folk stereotypes. If race is going to be assigned at all, it has to come from cultural evidence, not bones.

This tangle won't be easily undone. It's difficult to escape the divisions that anthropology and its predecessors set up—the language and ideas surrounding race—even as we recognize that there's no biological backing for them. Even distinctions such as Native American are recent inventions and not reflective of biology. There's no amount of DNA or anatomy that can make someone Native American. That identity instead resides in a culture that chooses you back and involves participation. As Pawnee historian Roger Echo-Hawk and anthropologist Larry Zimmerman write, "The Ancient One had lived and died without race. He entered the future as a prize to be won upon the racial battlegrounds of America." How to navigate past the old labels is a continuing challenge, and one that confounded the case of the Ancient One.

Such considerations of how to move beyond the racialism of anthropology were highlighted by the pain of legal proceedings that played out over two decades. The litigious anthropologists—as well as the court system—rejected the idea that the Ancient One's ancestry was connected to the Confederated Tribes of the Colville Reservation through their stories, and in 2002 the Oregon district court ruled in favor of the scientists. It was yet another case of Native Americans being told the nature of their own history, a continuation of a trend that had been in place for centuries, particularly during the height of anthropological and archaeological collecting in the nineteenth century.

In 1868, the surgeon general put out a call for US Army field officers to send whatever Native American skeletons they could to the Army Medical Museum. The number of such remains—the number of *people*—that were stolen is stomach turning. In the four years after the order was given, one surgeon alone shipped forty-two Native American skeletons to the museum. Another surgeon in the Dakotas raided the grave of a young Sioux woman to cut off her head and shipped it back east as "a fine specimen." Less than two weeks later, the same doctor was at it again, sneaking around with two of his assistants to decapitate the body of a Native American "before he was cold in the grave." In short order, the museum accumulated more than eight hundred Native American skulls.

This kind of collecting went on for another century, and more. In 1971, construction workers were building a new highway near Glenwood, Iowa, when they plowed into a forgotten cemetery.

Archaeologists were called in to check out the site and pore over whatever remains were interred there. Within the historic burials, they found the remains of twenty-six people who were appraised as settlers of European descent. They were boxed up in caskets and given new burials. Based on objects found in the graves, archaeologists also identified the bodies of a Native American woman and her child. They were taken to the state archaeologist in Iowa City. Native Americans haven't forgotten this treatment. "I remember reading about two Choctaw women who died in a Mobile, Alabama hospital and whose crania were sent to the Smithsonian in 1869," Choctaw anthropologist Dorothy Lippert recalls. "It seemed to me that being Choctaw and being dead were the only reasons they were identified as museum specimens and I promptly ordered a medic alert bracelet that reads, 'Do not accession my remains into the collections of the NMNH.' The bracelet is only partially a joke."

The Kennewick case was yet another battle pitting scientists who wish to know the history of humans against Native Americans who fervently hold to their cultural traditions. The united Native American nations were not swayed by the argument that the Ancient One was some unknown form of Paleoamerican. They kept up their protest as the case wound its way through multiple appeals and as the bones came to be housed at Washington's Burke Museum for safekeeping. But it seemed the scientists had won out. In 2014, two of the most outspoken anthropologists who had sued for access to the Ancient One, Douglas Owsley and Richard Jantz, oversaw the publication of a massive book on everything they had learned from studying the remains, from the chemical traces of what this person ate to their hypothetical body mass and appearance. The Ancient

One had been a coastal wanderer, the book concluded, enjoying the ocean-side bounty of the Pacific Northwest. But on the most critical question of whom the Ancient One was most closely related to, Owsley and Jantz drew a question mark. Throwing their support behind craniofacial measurements, they drew a line between the Ancient One—as well as some other ancient American remains lumped in under the title Paleoamerican to try and set them apart—and Native Americans. They couldn't conclusively say who the Ancient One was most closely related to or had come from beyond broad strokes, but they remained adamant this person was not Native American. Otherwise, their entire case for studying the skeleton would have vanished and the bones would have to be repatriated.

Then the stalemate broke.

It's rare that a single study has a major effect on science. Fields are so specialized, and hypotheses so narrow, that most major claims are treated as controversial hypotheses that require further testing and verification. A single paper in *Nature* or *Science* might spark wide coverage and discussion, but it's rare for any one piece of research to cause an instant shift. Yet, one year after the tome focused on the embattled skeleton was published, a genetic analysis instantly altered the Ancient One's backstory. At least, it did for scientists. For Native Americans, the news was already expected.

Previous attempts to subject the Ancient One's bones to DNA analysis had failed. Now, as it appeared in *Nature*, geneticist Morten Rasmussen and his colleagues had managed to extract and analyze the skeleton's DNA. What they found contradicted the idea that the Ancient One was a nomad from Asia or had deep European ties. Instead, they found the Ancient One has closer kinship to Native

Americans than to any other group. More specifically, they detected a genetic connection to the Confederated Tribes of the Colville Reservation, who were suing to have the Ancient One returned. Rasmussen and his colleagues undercut the idea that craniofacial measurements could be used to identify whom the Ancient One was related to, as well. Comparing populations to populations was one thing, but trying to fit a single skull into broader categories was inherently fraught—skull shapes are too variable. It's also possible for people to have a particular skull shape but an unexpected genetic makeup. The Arikara of North Dakota, for example, have skull shapes that might be classified as closer to Polynesians but their DNA and culture allies them with other Native Americans.

The news immediately ignited speculation about what might happen to the Ancient One. The geneticists were confident of their results, but some anthropologists—like Owsley—insisted that the Ancient One's skull proved a different kinship. He continued to call the Ancient One "a traveler," certain that "His people were coming from somewhere else" and that they belonged to some unknown culture. Other commentators argued the limited Native American DNA data set made associating the Ancient One with a particular tribe problematic. But the Confederated Tribes of the Colville Reservation had to think carefully about whether to let their DNA be sampled for comparison to the Ancient One. "Because of the way science has treated our people in the past, it was a tough decision," said tribal council chair Jim Boyd. They ultimately agreed to participate, and the genetic evidence provided the grounds to bring the Ancient One home. In December 2016, President Barack Obama signed legislation to repatriate the Ancient One to the Native American nations who had filed suit. They buried the skeleton in short order.

The Ancient One can finally rest, but that decision came at a great cost. It didn't have to be this way. Traditionally, Vine Deloria writes, "Archaeology has always been dominated by those who waved 'science' in front of us like an inexhaustible credit card, and we have deferred to them—believing that they represent the discipline in an objective and unbiased manner." The Kennewick case was an example that this exploitation continued. But there have been other incidents—while the Kennewick controversy was unfolding and since then—that serve as examples of how anthropologists and the people they wish to study can work together to treat the departed with respect.

In 1989, years before the Ancient One was uncovered, a Native American skeleton was found near Buhl, Idaho. It seemed to be very old, but radiocarbon dates were needed for confirmation. So archaeologists requested permission from the Shoshone-Bannock Tribes of Fort Hall to take a sample from the skeleton's humerus to get a date. The results put the age of the bones at about 10,675 years. Then, in 1991, researchers asked the tribe for permission to carry out additional analysis and make casts of the bones. This, too, was approved, with the corollary that no more bone was to be destroyed in the process. By the end of that year, following the studies, the skeleton was repatriated and buried.

The controversy and embarrassment over the Ancient One has only given researchers more reason to think carefully about unidentified remains and reaching out to Native peoples. We're still in a period of intense growing pains in which ethics are being examined, debated, and modified. Anthropology and archaeology have a terrible legacy, one that cannot be ignored, but there is still hope that the sins of this past can be surmounted and greater collaboration can teach us more than we would have otherwise known. The new model, researchers Chip Colwell and Stephen Nash suggest, should use a framework of informed consent when dealing with human skeletons with careful thought given to the wishes of the dead and their relatives. This doesn't necessarily mean every set of remains will be repatriated or reburied, but the point is to make an honest effort to contact tribes and ask about their wishes for remains of people related to them. Science can no longer assume they have permission to treat the dead however they please without consulting the people with ties to the deceased. And these considerations have started to trickle through into policy. The original NAGPRA provisions didn't consider ancient, unaffiliated remains. That's what let the Kennewick case explode and cause so much grief. But on May 14, 2010, a new regulation—43 CFR 10.11, or the Regulations for the Disposition of Culturally Unidentified Human Remains—attempted to patch that gap by covering remains assessed as Native American but without cultural affiliation. At museums around the country, anthropologists and archaeologists took the initiative to find a proper resting place for some of the skeletal unknowns in their collections.

Shortly after the legislation was passed, while the back-and-forth over the Ancient One continued, experts at the Burke Museum and University of Washington went through their collection to look for culturally unidentifiable Native American remains. These were skeletons with little or no information about where they came from, and the experts found dozens in their cabinets. As recounted by anthropologist Megon Noble, the two institutions got a grant to consult with the state's tribes about the remains and organized face-to-face meetings to discuss what should happen next. The remains that had some locality information were relatively easy to handle. Those were considered in terms of the people who had connections to that area. For the remains that had absolutely no information, the decision was trickier.

According to the law, the researchers technically didn't have to return the remains that lacked geographic backstories. But they decided to, anyway. A skeleton without its context is practically useless to researchers and, Noble points out, no one was studying them. "In 13 years at the Burke Museum, there were no research requests pursued on culturally unidentifiable remains with unknown provenance." There was no reason to keep the bones. In a conference call with nineteen attendees and representatives of fifteen tribes, Noble reported, the consensus was there would be no further investigation of the bones, but that repatriation was the goal. In July 2013, seven tribes gathered together on the Central Washington Plateau to bury fifty-three sets of unidentified Native American remains in the middle of the state, their chosen last resting place.

These discussions don't just concern Native Americans. The injury to them and their culture has been awful, but the renewed emphasis on collaboration and consent has also led anthropologists and archaeologists to question what should be done about non–Native American remains that came from suspect or unknown sources. Colwell and Nash tried to provide a model for what should happen to bones of unknown origin and affiliation. After locating non-Native mystery remains in the collection of the Denver Museum of

Nature and Science, they held an interfaith meeting with museum staff and representatives of various religions. It was decided that the bones should be respectfully buried, and they were scanned ahead of time to make sure no information would be lost to science in case some later investigation wanted to look at them. This raises questions about what should happen to bones in other osteological collections that were stolen or acquired without permission. At what point do we stop looking at someone as an exhibit and restore their humanity? This isn't hypothetical. That question directly applies to skeletons like that of Charles Byrne.

During the late eighteenth century Byrne became famous as the Irish Giant, standing about seven feet seven inches tall. He was heralded as a gentle soul, lauded for his charm as much as for the stature people gawked at. But Byrne also knew some people were looking at him with a sharper gaze. This was the heyday of medical experimentation and specimen collecting in London. Byrne knew that after death he would be turned into a collectible. So he made express wishes to be placed in a lead coffin and sunk in the ocean so that he wouldn't wind up on the dissection table. But a surgeon named John Hunter would not be dissuaded. When Byrne died at twenty-two from complications related to a then-unknown disease called acromegaly (the cause of his famous proportions), Hunter arranged the theft of Byrne's body. Byrne's skeleton was soon displayed at the Royal College of Surgeons, where it has remained ever since. There seems to be little willpower on the part of even modern medical experts to return the ill-gotten bones. When, in 2011, ethicist Len Doyal and law professor Thomas Muinzer renewed calls to do right by Byrne and bury him according to his wishes, there was reinvigorated public support for honoring what Byrne had intended. The Royal College of Surgeons considered the matter, and decided not only to keep Byrne, but to keep him on exhibit as the centerpiece of the Hunterian museum, where he draws visitors to their corner of London. Another effort to rescue Byrne was mounted in the spring of 2017, but the British doctors have remained steadfast. They say there is still much to be learned from Byrne's skeleton, though what they are principally demonstrating is how cruel we can be to the dead.

The relevant and macabre cases aren't just old news. It's still happening, and body theft has become quite a moneymaker. Since 1995 the touring exhibit Body Worlds has been presented as a unique and detailed expression of what's inside us. The plastination technique developed by German artist Gunther von Hagens allowed cadavers—or parts of them, in some cases—to be presented as if they were still alive, offering visitors a greater appreciation for what lives in us but goes unseen. Yet, despite von Hagens's repeated assurance that he obtained his cadavers ethically and with consent, in 2004 proof emerged that at least some of the Body Worlds cadavers were executed Chinese prisoners. Bullet holes and an email chain left little doubt that the bodies were illegally obtained; at least seven had to be sent back to China for burial. Competing exhibitions—such as Bodies Revealed—have faced the same charge, with any bodies obtained from China, Russia, and Eastern Europe facing immediate suspicion as to their origins.

All of this might seem like high-profile intrigue, smuggling the bodies of prisoners for exhibition seeming like a twisted plot for a pulp thriller. But the trade in bodies and body parts runs both broad and deep, bustling along in plain sight. This is what journalist Scott Carney calls the Red Market, which offers everything from kidneys for transplant to bones for decoration.

From the name alone, the Red Market sounds like a shadowy, secret alley in a game like *Skyrim* where cloaked figures offer rare, illicit, and illegal goods. But the truth is not really hidden at all. Trade in stolen bones and skeletons often takes place out in the open. In 2007, for example, authorities in Calcutta, India, raided a medical supply company called Young Brothers. Neighbors had complained about the horrible stench emanating from the storefront, and workers there sometimes laid bones out on the rooftop to dry. The company had been buying these bones from thieves who stole bodies from graveyards, Hindu funeral pyres, and rivers, purchasing skeletons for resale at about $45 apiece. Law enforcement found literal truckloads full of human bones as well as documentation that human remains were being shipped to the United States, Singapore, and other countries despite a ban on bone exports that had been in place since 1982. Young Brothers' owner, Vinesh Aron, was promptly arrested but soon released. There seemed to be little will to stop the business, and Young Brothers still maintains a web sales page offering "HUMAN SKELETON MATERIAL (ORIGINAL BONE) for your kind consideration & order."

Why India has become such a hub for grave robbing and illicit bones takes us again back to the nineteenth century. During this time, when India was part of the British Empire, the English medical school boom caused a crisis in the UK. There was such a demand for cadavers that people looking to turn a quick buck raided graveyards to supply the academics. In 1828, the infamous pair of William Burke and

William Hare went so far as to murder sixteen people to get fresh bodies for Dr. Robert Knox's medical classes. These "Anatomy Murders" shocked the public and politicians alike, culminating in the Anatomy Act of 1832, which tried to curb the rampant crime surrounding the medical schools of the time by giving doctors and students access to the unclaimed bodies of those who died in hospitals, prisons, and workhouses, as well as those who donated their corpses for dissection. But this wasn't a perfect patch. People still protested that the poor and destitute were being cut up without consent, and the demand for bodies still outstripped the legal supply—in England as well as the United States. So the trade shifted to India, where it was easy to snatch bodies off the ceremonial cremation pyres at the center of Hindu funeral rites and drag out bodies that had been deposited in holy river waters as part of standard burial practices. This continued unabated for more than a century, until 1985, when one bone seller was reportedly found to possess more than fifteen hundred skeletons of children. India quickly banned the export of human skeletons in the wake of the news, but the regulations have done little to stopper the market. Bones from people who perished in India still make it to other countries—particularly those that don't ban the import of human skeletons—and there has been a steady demand among India's own medical schools for real human bones.

Just like England and the United States in the nineteenth century, India has been undergoing its own medical school boom, and students are told to learn directly from real human bones. Their teachers insist that plastic casts or generalized replicas are not detailed enough to learn from—which is not true when the aim is to learn generalized human anatomy—and so each medical student is encouraged to purchase their own set of human remains. The market even within the country remains huge; in 2017, eight people were arrested on suspicion of pulling bodies from West Bengal's rivers to skeletonize and sell. Desperation drives the commerce. People who need money quickly—and who are often in impoverished communities, whether it's in India, Burundi, China, or elsewhere—will often sell their kidneys, hair, or blood to get out of their dire situation, and the body brokers certainly don't argue with them. Each arrest or prosecution ends up being just a small strike back at a massive industry that shows little sign of stopping.

That's hardly all. The human bone trade is as close as the phone in your pocket or the laptop on your desk. That's because the sale, purchase, and trade of most human bones in the United States is largely legal. There are some stipulations. Native American remains are protected under NAGPRA, and each state has different provisions about how human remains on the market may be obtained or carried over state lines, but, provided you have the cash, you can order a freshly cleaned human skull or an "antique" skeleton in minutes. There's no federal regulation or oversight for such private sales. The online store for a Los Alamitos, California, shop simply called the Bone Room assures buyers, "it is perfectly legal to possess and sell human bones in the United States." Click a link on the same page for human skulls and you'll likely find listings like "#6043 India male," a slightly damaged skull going for $1,800. The listing doesn't provide any information on who this person was, how the skull was obtained, or even if it came to the United States before India's 1985 export ban. And therein lies the troubling nature of the casual, open trade in human bones going on every day.

Bone gathering has become a subculture status symbol. Private collector Ryan Matthew Cohn boasts more than two hundred human skulls in his personal collection, no doubt adding to the required mystique for his duties as the host of the Science Channel show *Oddities*. Artist Zane Wylie has also made a name for himself by purchasing and artistically modifying real human skulls. People going for a Tim Burton–esque aesthetic or wanting to gain some goth cred can easily acquire and display human remains to set the appropriately macabre mood, with exotic, old, or unusual skeletons being the most desirable status symbols of all. Find and follow the right hashtags on social media and the Red Market is easy to navigate.

Instagram, which more readily brings to mind carefully framed selfies and a flood of tattoo art, might not initially seem to be a likely hub for the human bone trade, but that's changed in recent years. Part of it's because other popular websites and apps have tried to ban the sale of human remains. The craft-centric Etsy disallowed the sale of human bones in 2012. Similarly, eBay, which was much more permissive, put a ban on all human body parts except hair in 2016. In this case it wasn't so much out of a newfound sense of virtue as a scientific report tracking the auction site's market for human skulls. Christine Halling and Ryan Seidemann of the Louisiana Department of Justice tracked the sale of 454 human skulls on the site, noting that fifty-six of those were "of forensic or archaeological interest" and shouldn't have been listed for sale. This came as no surprise—in 2009 Seidemann and colleagues reported on a Native American skull that had been offered on eBay and subsequently seized by the State of Louisiana—but the conclusions of the new paper were stark enough that eBay changed their store policy within a week and barred the sale of human bones. So the market moved to Instagram, where experts are tracking how bones are shuffled around.

Archaeologists such as Damien Huffer track how human remains are marketed and sold over people's smartphones, and, with colleague Shawn Graham, he dug into the mechanics of the trade in a 2017 paper called "The Insta-Dead: The Rhetoric of the Human Remains Trade on Instagram." As it turned out, the language used to promote and purchase skeletons is very familiar. The general vibe is that of nineteenth-century archaeology and anthropology, putting the acquisition of specimens above the recognition of those pieces as people. "In the same way those early collecting practices did damage and violence to communities from which the dead were collected," Huffer and Graham write, "the emergence of social media platforms that facilitate collector communities seems to be re-playing that history." The bones of these people are effectively stripped of their humanity—with little to no information about who they were, where they came from, or how the remains were obtained in the first place—to become, simply, *things*. Hand bones become the basis for artsy necklaces and a skull under glass is a coffee table centerpiece. It's another way that people are turned into objects, and it seeps into the language of the science, too. "The ability to sell, display or trade human remains via social media and online distribution lists has led to their being treated as consumer products for a collector's market," Huffer and Graham write, "rather than objects of archaeological, ethnographic or anatomical value."

As it stands now, the bone trade in the United States is largely legal. But not all of it is so irresponsible. One particular purveyor—Skulls Unlimited International—gets its fresh skulls from donors, and these are sold to doctors or researchers with a particular need for real remains. But there's still an untold number of bones and skeletons of unknown origin being swapped on Instagram—and according to Huffer and Graham, even Etsy, despite the ban—because they are labeled as antique or historic. These are older, broken, stained remains that are often said to come from estate sales or deaccessioned medical school collections. They are one step removed from today's active bone market, and perhaps purchasers feel absolved by this. But a skull said to have come from an old medical school collection is still likely to have been originally obtained in an unethical way. While major bone dealers claim they can spot bones that have been robbed from graves or are otherwise sketchy, there's no such filter on Instagram. Regulatory bodies—whether it's social media sites or even law enforcement agencies—have been slow to regulate these sales.

The market continues to grow. In their study, Huffer and Graham tracked attempted sales of human bones on Instagram between 2013 and 2016. While there were only three relevant posts in 2013—asking prices totaling $5,200—there were 77 in 2016, with asking prices coming to a total of about $57,000. For the most part, these aren't wealthy rarities dealers. They're oddities collectors, artists, and part-time dealers who buy and trade on the small scale, using tags like #trophyskulls and #realbone to market their offerings.

The trade makes me shudder. I wouldn't mind winding up as a skeleton in a museum cabinet after I die, teaching even after death, but the thought of being bought and sold as a decoration, appraised to a certain value, winnowed down to nothing more than an object of curiosity gathering dust, makes my living flesh crawl. Treating human bones as a curio divorces you from history and context. Bone, after all, is the most lasting part of ourselves, able to speak to succeeding generations even after our voices have gone silent.

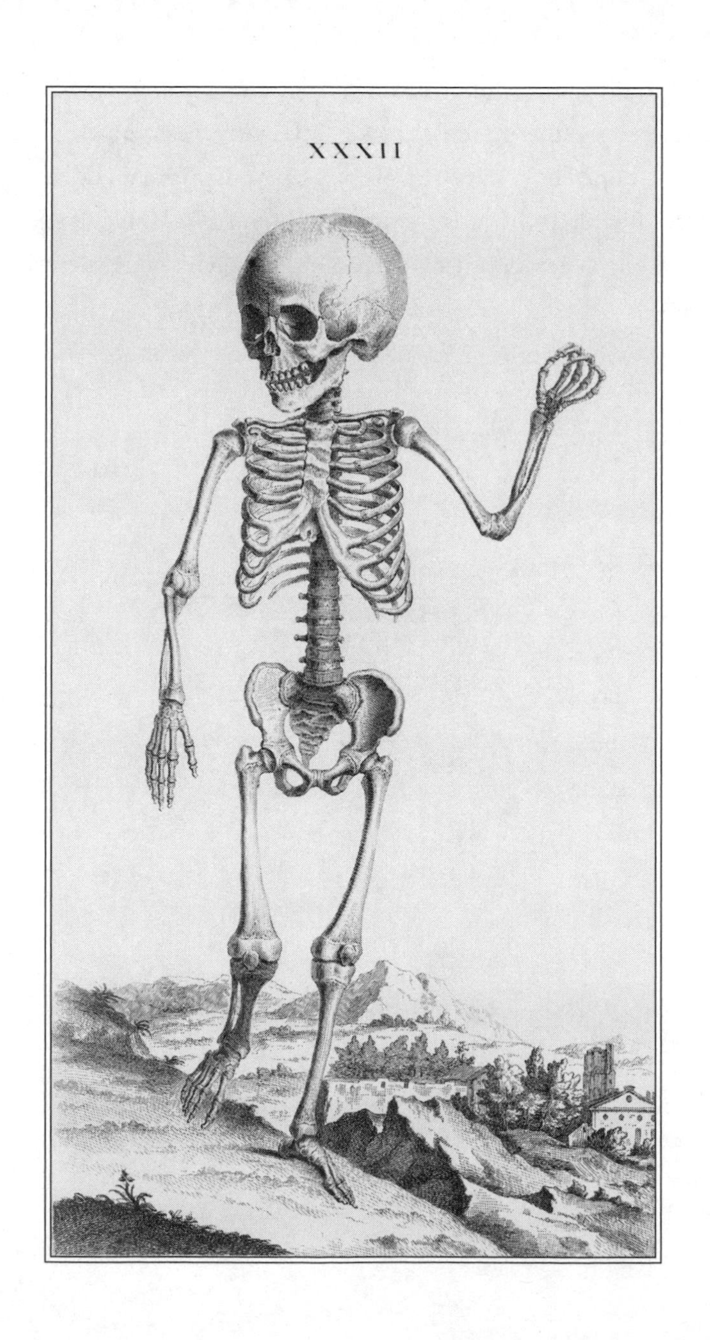
XXXII

TEN

BONE DEEP

Sometimes I like to just be still and think about my bones.

It's something I regularly do when I'm bored and trying to make the hours melt away. The last time was while I was hiding from a torrential downpour in Utah's eastern desert. To be fair, bones were already on my mind. I had spent days toiling away in the 105-degree heat—nothing like fieldwork in July—on the petrified bed of a 157-million-year-old dinosaur, the skeletal pieces only stubbornly letting us take them from the rock. It was slow work, the passage of time tracked by the amount of stone chipped away from the maroon-colored bones and the gradual thunderhead buildup over the distant Abajo Mountains. Now and then, those storms would visit and give the crew an excuse to huddle in a shallow sandstone cave downhill from the elevated quarry, trying to forget the fact that lightning could still strike us there.

During these forced breaks, most of the crew would close their eyes and start to doze. Rhythmic snores lulled those who hadn't already drifted. But I couldn't sleep. Relaxation has never come easily to me. Instead, my arms folded behind my head and the tips of my boots misted by the downpour at the edge of the overhang, I thought about my skeleton. If I were to be totally stripped of all my flesh and viscera, but still kept alive by some kind of magic, what would I look like as I lay there? An X-ray version of myself, each joint shifting and flexing as I tried to get comfortable and as I simply breathed, my rib cage slightly expanding and falling back even as I tried to remain completely still. Would anyone be able to tell it was me? Maybe. Once, while at a conference in Washington, D.C., an osteologist acquaintance of mine walked up from behind and said, "I knew it was you from the shape of your skull!" It was an odd sensation to try to focus on my bones—not so much an out-of-body experience as an inner one, trying to envision each of the two-hundred-some-odd parts in their place.

Try the meditation yourself sometime. The next time you're waiting for a flight or for a movie to start, or if you can tear your eyes away from your smartphone in a moment of quiet, think about your bones. Concentrate on what's beneath the surface, what you can sense but cannot see. Hands are wonderful for this. They're the most mobile parts of our ape skeletons, after all, and among the most personal. Hands are how we experience so much of the world around us, and they carry more character than we often realize. And inside the rind of skin, muscle, and ligament are stacks of flat, fiddly little bones, connected to your lower arm by a gap that ends up making a flexible hinge. Then try it with the other parts. There's a spine inside you. There are all those skull bones, biologically welded together just below the surface of your skin. Trying to envision what a navicular or cuneiform is doing at any one moment is probably going too far in, but you get the idea. Envision your skeleton by itself for a moment, the core of who you are.

But that's only considering the skeleton as a fact of nature, a manifestation of what *is*. What all those bones mean depends on your point of view. When I think of bones, I think of Darwin's "endless forms most beautiful and most wonderful." Everything about the bones inside us, from their arrangement to their microscopic structure, is a testament to the way evolution mixes blind chance with the winnowing edge of natural selection. By mixing and matching old parts, forced along only by what's useful in any given moment, what's old becomes something new. But that's hardly all. We carry the past in our bones. Our species is relatively young, still a long way away from the million-year average that most mammals tend to persist, but even though we prefer to believe in our novelty, our skeletons underscore the truth. The basics of our body plans were forged in the seas in a series of happenstances, with tweaks and refinements coming from life on land and in the trees. Our evolution continues, but we're mostly able to pick out these differences because we've developed a talent for noticing patterns in our own species. From the broader view of the fossil record, there is nothing about you or me that is particularly unexpected or staggering. We're variations on a theme, a new combination of features that makes us stand out but also, more important, joins us to a history longer than any of us have a hope of truly and fully comprehending.

I can only imagine what a future intelligence—our descendants? life from elsewhere? another species that happens to evolve the wisdom to examine its past?—would make of us, or at least those of us who leave our bones to the fossil record. It's really the best chance we have at lasting beyond ourselves. The legacies that we try to build are either dimmed or destroyed by the passing of time. There is almost nothing that we can create that holds any permanence. But if summer after summer of scuffing my boots on arid rocks and feeling the back of my neck crisp as I scan the ground has taught me anything, it's that bones are the one chance we have at lasting millions upon millions of years, the purest and most minimalistic records of who we were. Even better, we don't have to wait for happenstance. With a little forethought, and hopefully someone willing to carry out our wishes, we can become fossils.

The thought first occurred to me while hiking alone down Park Avenue of Utah's Arches National Park one June afternoon. There were no towering skyscrapers here, but from the tall stone walls, it was easy to see how the short trail got its name. And even though it was not remote by any means—you can stand at one end and see the park traffic go by at the other—the orange and rust sandstone provided that most essential of desert comforts, shade. It was peak season, but I hardly saw another person as I ambled along the slickrock below, a few croaking ravens perched in the nooks of the Jurassic rock above being my main company. And after I turned around and started to make my way back, I stopped to look at the sandal prints I'd left behind in a few dry pools of rust-colored sand. How long would they remain there? Would they have any chance of withstanding the ages, like the dinosaur tracks that pock the stone at various places around the park? Not likely. If they weren't brushed over by another tourist, the wind or occasional thundershower would wipe them away, not to mention that this desert was an erosional environment—a place where the elements were chipping rock away and moving it elsewhere, not laying it down to be preserved in perpetuity. But the petrified cogs of my mind kept whirring as I climbed the trail back to the road. In slightly different circumstances, those prints might have been preserved for a time as deep as the surrounding rock walls. The fossil record isn't something of the past, but grows every single day existence keeps rolling on. If I were to become a fossil, how would I want to do it?

Fossil is not synonymous with bone. Footprints can be fossils. In fact they're sometimes more informative about the way an animal lived than bones are, given that traces are actual moments preserved in stone, like the trail at Laetoli. I could pick various mudflats and lakeshores, walking back and forth barefoot to leave my tracks behind, and if I'm lucky, some of those might dry up and harden only to be buried and preserved by the next wave of incoming sediment. (Or if I really wanted to confuse paleontologists of the future, I could leave my sandals on, letting them wonder what "Vibram" means.) But the thought of tracks being my permanent record doesn't appeal to me very much. All the future would know of me would be the soles of my feet and, with the right calculations, my height, walking speed, and the fact that my feet tend to turn outward as I go along. Nor was I very happy with the contributions to the fossil record I've already made. Like billions of others, I have generated plenty of garbage that's rotting in trash heaps and driven vehicles that have belched a horrific amount of greenhouse gases into the air, contributing to the biological crisis that may mark this time in history not so much as an era but as a mass extinction event. I don't want my legacy to be a break of barren rock that marks the latest of the worst die-offs in history. Bone has to be the way to go, and here a science called taphonomy will be our guide.

Even though it didn't have a name yet, taphonomy got its start with the help of the eccentric British clergyman William Buckland. Buckland was off the mark with his identification of the "Red Lady," but his main claim to fame was that he founded the study of how fossils are made. This was his work at Kirkdale Cave in Yorkshire.

In 1821, local quarry workers found a cavern with a vast jumble of bones buried in its floor. Laborers, amateur collectors, and local parish heads all descended on the spot, plucking up mementos from this place that was said to be paved with osteological treasures. Early identifications suggested a mix of animals—mammoths and rhinos as well as foxes and abundant hyena bones—and this news puzzled Buckland. Deposits like this were supposed to come in one of two flavors. There were fissures that the bones of long-lost herbivores were swept into—a phenomenon Buckland attributed to the "Noachian deluge"—or caves that carnivorous mammals used as dens. To have an abundance of both kinds of remains didn't seem to make any sense. So, despite the winter chill, Buckland crawled into the cave himself, and even though collectors had already been messing around in the cramped space, he nevertheless was able to determine that there was no fissure for animals to tumble in through. They must have been dragged in here by the voracious hyenas at a time that, from the cave's geology and a Christian faith that did not yet have to reconcile with the reality of millions of years of evolutionary change, Buckland put at just before the great flood.

But it's one thing to come up with a story and another to test it.

That's what science requires—the persistent but essential little gremlin that whispers, "Is this testable?" when you think you've come up with a brilliant solution to a problem. Buckland did just that. Among the fossils previous collectors had overlooked was something that Buckland had already taken a keen interest in—prehistoric poop. He plucked up a few of these plops, suspecting that they had been left by the cave's hyenas, and, sure enough, his chemist friend, William Wollaston, confirmed that the scats had exactly the high bone content you'd expect. Buckland even went so far as to ask France's Georges Cuvier, the most respected anatomist of his or perhaps any era, to send him the crap of a hyena that lived at the Museum of Paris, and these comparisons, as historian Martin Rudwick wrote, "clinched the case."

But Buckland did something else that was just as critical. After he returned to Oxford, the implications of the cave for connecting the past world to the present buzzing in his skull, a traveling show with a spotted hyena passed through town. Buckland offered the beast a selection of ox bones and watched carefully which ones the hyena plucked up, how it broke them open, and, eventually, what came out the other end. It turned out to be a near-perfect replay of what must have happened at Kirkdale; the pattern of breakage and gnawing was practically identical to the fossil bones from the cave. Modern hyenas had bridged the gap between the world as we know it and what came before, even explaining their part in forming the fossil record by bringing bones into a place where they would eventually be covered.

You can still see some of those experimental bones today in a quiet little corner of the churchlike Oxford University Museum of Natural

History. These cracked remains are behind a glass pane with a few fossilized and more recently gnawed bones side by side. They're beautiful, despite the bone-crushing violence that created them, and I wanted to run around to the quiet families gazing at the hall's dinosaur skeletons and drag them over to the darkened corner to show them the bones that launched a science. I held back from doing so—everyone knows that unless you're a paleontologist or archaeologist yourself, a strange man insisting that you look at old bones is how horror movies start—but, really, I just wanted someone to share my joy as I fawned over the battered fragments propped up behind the glass. They weren't just hyena leftovers, but proof of the geological maxim Buckland's student Charles Lyell would eventually coin—"The present is the key to the past."

The reaction to Buckland's "hyena story" was momentous. Even if his colleagues looked down their noses at his methods—what distinguished professor wrote letters to acquire fresh poop?—they could not argue with his results, particularly as he attempted to place Kirkdale Cave in the context of how the world had changed. Buckland even won the highest honor available to geologists, the Copley Medal, for this work. That's why it's strange that his interest in reconstructing prehistoric events did not catch on among his peers. Maybe it was too dirty for the respectable men of science. Perhaps fieldwork, crawling through caves and feeding carnivores the butcher's leftovers, did not appeal to anatomists, who preferred the cleanliness and order of the museum lab and writing desk. Or maybe it was because there was so much novelty in the fossil record that simply describing the various pieces that had been found and how they fit together was a job bigger than any scientist could hope to accomplish in their lifetime. Especially when the badlands of the American West were found to spill out an abundance greater than anything ever seen in Europe.

Still, the bigger point of any study of prehistory is to put the past in its place against the watermark of the present, perhaps even joining the two. As much as I love the phrase "lost worlds," the fact is that it has always been the same world, with today's life inextricably entwined with that of the past. Processes that occur now did not just pop into existence for us to observe them—they've been going on as long as there has been life.

A German naturalist who wandered around the Gulf Coast of Texas made a forceful argument for this science. Johannes Weigelt articulated his view in *Recent Vertebrate Carcasses and Their Paleobiological Implications*. It's a journal, a thesis, and certain to put off the squeamish. The title page of the translation in my library shows dozens of skeletons laid out in a fetid pool, tatters of flesh still clinging to their bones. "Carcass assemblage of horses that died of starvation, near Kraslawka, west of Dunaburg," the caption reads, identifying them as forgotten victims of the Russian Revolution. The photos at the back of the book present more of the same—not carnage, but snapshots of the afterlife many vertebrates face. After clean line drawings of various fossil creatures are photos of cows bloated with gas, desiccated fish, an alligator in increasing states of decomposition, beached dolphins, and more. These were the bodies that instructed Johannes Weigelt about how fossils are made—the carcasses both embodying the fossil record of the future and acting as natural experiments revealing what must have been happening for more than 508 million years.

Published in 1927, the book was the culmination of sixteen months of studying the afterlives of animals in Louisiana, Oklahoma, and Texas, not focusing on what was in the meager rock outcrops of the area but looking to modern animals to reflect the past, seeking answers to the questions that tugged at Weigelt's brain when he visited museum collections. "How did all these animals die?" he asks in the book. "What happened to them before they were embedded? What particular conditions enabled their preservation in such great numbers?" It turned out that decomposition and preservation were dynamic happenstances; bodies had entire afterlives influenced by everything from season of death to speed of burial, each set of remains in the fossil record a unique expression of happenstances. To that end, Weigelt surveyed everything from mode of death—ranging from explosive volcanic eruptions to quicksand and "Death on Ice"—to what happens to a body exposed to the elements for varying lengths of time before burial.

All of this, in time, established taphonomy as the science between life and death, or what paleontologists often sum up as "what happens between death and discovery." Pioneers in the field such as Anna Kay Behrensmeyer continued the investigation, following what the breakdown of a body can say about that creature and the environment it inhabited. Read in reverse chronological order, these phenomena can rearrange a skeleton in our minds up to the moment of burial, if not death. And once there, we can also go forward, back up through time, sifting for clues as to how this might work for ourselves.

First of all, it's worth considering what *can* become fossilized. Death is merciless about paring away information, little by little.

DNA starts to unspool right after death, breaking into smaller and smaller tatters with time. Even under ideal conditions, like if a creature comes to rest in a cool, dark cave, the genes inside its cells break down and leave only traces of what they once were. In fact, DNA held within bones breaks down at a fairly regular rate, much like how radioactive minerals gradually change over to inert components, having a half-life of about 521 years. This means that every half a millennium there's half as much DNA as the bone started with at the beginning of the count under even ideal conditions, with the basic math meaning that the entirety of a bone's genetic material will decay away within about six million years. That's a long time compared to a human life span, but it's still pretty quick. So while bones can and do contain DNA for a long time after death, our genes—like all our other soft parts—are ephemeral, winnowed away day by day until there's nothing left of them. And thinking long-term, this means there's no hope of saying "Bingo! Dino DNA!" while studying *T. rex* or its Mesozoic relatives. Genetic material simply doesn't last long enough, meaning birds are the closest we're ever going to get to seeing a *Velociraptor* running around. All the more reason to watch ravens with admiration and a respectable distance.

There are other, more abstract things that have a hard time entering the fossil record, too. Intelligence, for one. Skulls and natural casts of brains can show us the anatomy of cerebral control centers and their size, but that tells us very little about the intelligence embodied in that soft tissue. Likewise, sound is something that can only be drawn from the fossil record under just the right circumstances. Even though there are at least eleven beautifully preserved skeletons of *Archaeopteryx*, the first bird, that include everything from rings of bone in the dinosaur's eyes down to its feathers, we know nothing of the tissues in its throat. Even if we had them, we'd be at a loss to figure out how they made sound and conclusively determine whether the first bird sang, croaked, hissed, or stayed silent. Sound only survives when it's made by structure. Such was the case with a 165-million-year-old katydid named *Archaboilus musicus*, which chirped by rubbing a bumpy edge of a wing against a scraper on the other. Because of the fossil's intricate preservation, paleontologists were able to reconstruct what the Jurassic insect sounded like. Unfortunately, not being an arthropod, I don't have any equivalent sound-making structures, and so I won't leave any record of what I sounded like with my body.

Color presents a similar problem. Paleontologists are able to reconstruct fossil colors in exceptional cases when they find tiny little blobs called melanosomes intact in feathers, fur, and other body coverings. These globules create color structurally, scattering back light in particular parts of the spectrum, from rust red to iridescent black, based on their distribution and density. The fossils themselves no longer show these colors—most often, they look charcoal gray to our modern eyes—but comparing the distribution of melanosomes in the ancient feathers to those of modern birds with known color patterns allows the true colors of Mesozoic dinosaurs to shine through. Melanosomes have even been recovered in the scaly skin and armor of Cretaceous dinosaurs. Whether someone would be able to reconstruct my personal color palette or not, I couldn't say, though I know it wouldn't be as impressive as the scarlet spikes of the armored *Borealopelta* or as useful as the countershaded camouflage of the little horned dinosaur *Psittacosaurus*.

The upshot is that the preservation is often a process of loss. It's just a matter of where that process stops. Start with any vertebrate body and various details will likely be torn off by scavengers, disrupted by bone-burrowing insects, and degraded by bacteria, along with the cracking and weathering that comes with exposure to the elements. When you look at everything that's constantly working to break down the shape of life, it's amazing that we have any fossils at all.

Still, the fossil record doesn't always operate by what we might think of as the most logical pathways to osteological immortality. Sometimes the very phenomena that could totally destroy a skeleton end up saving parts that would be lost to us otherwise. For example, being eaten by a predator. Fossil scat often records evidence of what a creature was eating, and for carnivorous mammals and dinosaurs this means that bone fragments sometimes wind up coming out the other end enclosed in a steaming pile that helps shield them from the elements.

Even the feeding habits of carnivores can have their advantages, as the human fossil record itself shows. Throughout our history carnivores have had a habit of taking us, or parts of us, back to places where we're more likely to be buried. Some of the 1.8-million-year-old hominin bones of Tanzania's famous Olduvai Gorge bear the bite marks of the local crocodile—named *Crocodylus anthropophagus* in honor of its meal choices—that may have dragged the unfortunate prehistoric people into a watery grave that was more likely to cover up their bones than dry land. Another fossil from Africa—the roof of a hominin skull known as SK 54, found at Swartkrans Cave—is perforated by a pair of holes that match up with the lower canines of a leopard, and if the cat lived in or around the cave, it might have dragged the young australopithecine off to a quiet spot to consume it just as it would an antelope. And the famous *Homo erectus* bones accumulated at Zhoukoudian, China—noteworthy not only for their role in the development of paleoanthropology but also because most were lost without a trace at the outbreak of WWII and are now only known from casts—were likely brought into the cave by the giant hyena *Pachycrocuta*, the pattern of bite marks on the skulls and other bones showing how the carnivores deconstructed their human prey according to a regular pattern in order to get at muscle, tongue, and brain. In these cases and others, predators have actually granted us a look at our ancestors and close kin by unintentionally depositing their battered bones in protected places.

For my own sake, though, I'd prefer not to be snatched up by a crocodile or torn limb from limb by hyenas. Even after death, as science writer David Quammen has eloquently pointed out, there's something unsettling about the concept of being eaten. Given that I have all of the fossil record at my disposal for clues about how to increase my chances of entering it, I would prefer to give my bones the best possible shot at becoming preserved. And as taphonomic experiments have shown over and over again, that means extremely rapid burial. Bodies break down quickly when exposed, and as surrounding soft tissues decay, they expose bones that are bleached by the sun and made use of by a varied and efficient invertebrate cleanup crew. While out looking for fossil skeletons, I've more often run into what's left of modern animals that passed away some months or years before, the outer surfaces of their brittle bones peeling and cracking like a postmortem sunburn. You can even get a rough idea of how long those bones have been exposed by looking at how badly they've weathered. And at such a point the bones might not even survive their long geological interment, winding up as just a few friable shards. For wannabe fossils like myself, rapid burial means everything.

But what sort of environment to choose? There's no single setting that lends itself to paleontological perfection. For a few ideas, I pulled a copy of a book called *Exceptional Fossil Preservation* off the shelf and, though not the intent of the authors, used it as a kind of brochure for what I can hope will be a long, long afterlife. Right off the bat, a Burgess Shale–type burial looks pretty good. If it was good enough for *Pikaia*, it's good enough for me. But the trouble would be situating my body at just the right place at just the right time to not only be buried, but have the rest of the marine community buried by muddy silt so that my remains wouldn't be picked apart and scattered by the busy little fish and crustaceans that would treat me as a delicacy. Many of the other early options suffer similar problems of timing, not to mention that they preserve animals much, much smaller than me. At five foot ten inches tall I'm not a giant anything, but I'm still big enough that a significant amount of sediment is required for a proper blanket. On that score the Berlin-Ichthyosaur site in the middle-of-nowhere Nevada looks better. It contains the skeletons of enormous marine reptiles that died en masse more than two hundred million years ago, but, frustratingly, all that's known about the final moments of these fishlike saurians is that they were buried in deep water. There are no actionable instructions for how to repeat the process.

Osteno, Italy; Germany's Posidonia Shale; and the Oxford Clay of southern England—all places famous for their delicately preserved fossils—don't offer me much hope, either, other than reinforcing the impression that just about all of these fantastical fossil beds were formed in a marine environment where sediment was plentiful but whumpfed down on the resident species without a set schedule. Granted, there would remain some chance that I'd leave *something* to the fossil record in modern seabeds that experience irregular inundations of marine sediment—not to mention deserts where giant sand dunes regularly collapse, floodplains that get choked with new sediment, and, if there's a volcano nearby, calm lakes where ash falls in abundance—but what I'm looking for is the best possible option available. And that points me to something like Solnhofen.

Around 150 million years ago, at the time my beloved *Brontosaurus* was stomping around Jurassic Utah, the area around Solnhofen, Germany, was part of an archipelago scattered across a warm sea, spits of land enclosing lagoons along some stretches of shore. Little toothy pterosaurs fluttered through the air, horseshoe crabs crawled in the shallows, and fuzzy dinosaurs scampered around on the beach, and we know all this because of the stunning fossils preserved in the lithographic limestone that made this part of Bavaria famous. This is due to bad fortune on the part of the creatures but a fortuitous set of circumstances in the water. Storms swept across the islands in the Jurassic just as they pound other island archipelagos today, and when they did, the wind and waves washed creatures out to the depths of the lagoons. Some were already dead, others were unfortunate enough to be killed by the weather, but in either case their bodies tumbled through the water column to a bottom layer almost devoid of oxygen. We know this because there are almost no worm traces or other signs of activity. The lagoon bottom was hostile to life, and this was perfect for the dead. All those deceased species could rest without being hassled by scavengers or scattered into a million fragments. They truly rested in peace, rapid burial by falling sediment coating their bodies with such a light touch that even the scales and wispy protofeathers of the island's dinosaurs were left intact.

The Solnhofen method of inhumation sounds like as close to a sure thing as there is in preservation. There are always risks, of course. Rapid burial is still needed to keep delicate details intact and ensure that bones stay articulated, and then there's the fossilization process itself. The rock has to harden at the right time and mineral-carrying water has to percolate through those tissues in order to etch them into stone. Then those rocks have to be uplifted in just the right way, brought close to the surface so there's even a chance of discovery. There are so many things that can go awry. Still, if you wish to leave the ultimate record behind, you could do much worse than a long rest in the deathly calm at the bottom of a lagoon.

I don't know whether I'll wind up in such a place. I don't have a death wish to do so, and I'm not sure where I'd find an equivalent environment today that'd produce the desired result. Our current social and legal system generally frowns upon leaving your body wherever you please. But even if I take the shorter-term option of leaving my skeleton to science, I have to wonder what my skeleton would say about who I am and who we are as a species. There'd be no sign of what I was able to write or accomplish. No specific memories. All that would be left would be naked biology, an internal record that I managed to grow during my brief time here.

I wonder what a future paleontologist would be able to glean about my bones. Much of it would depend on how intact my skeleton would stay. Most every vertebrate paleontologist hopes for an articulated skeleton, will gladly accept an associated skeleton, and often curses a pile of fragments. But let's say for the sake of our argument that the scavengers aren't too harsh in taking what they require and that burial is fairly rapid, leaving me a bit of a puzzle but still clearly all belonging to one individual. If my hips remain intact, any future anatomist worth their salt will be able to figure out that I'm an osteological male. The flanges of bone at the base of my pelvis are narrow and could not have allowed a child to pass through. A histological section of my right radius might show that some of the bone there was remodeled when I was relatively young, the whisper of a greenstick fracture from a slip off a skateboard when I was ten. Examination of my teeth might show the little dent in my right upper incisor from where I used to habitually pop open raspberry pips before I noticed the little notch in the mirror one morning, and the microscopic scratches on my teeth—microwear—would show that I generally ate soft foods but had a penchant for cracking ice cubes and hard candies with my molars. And while my arms are a titch long and my hands are on the big side, I was of unremarkable height and anatomy, so standard that the future paleontologist who discovers me will probably be disappointed by my averageness.

Even if a scrap of me winds up preserved, be it in a museum cabinet or in the sea bottom, that will be enough. It's not that every skeleton has stories to tell, or even every bone. Every little shard has something to say about past life. Even a little piece would be enough to say a few things—that I was a vertebrate, that I was a mammal, that I lived right as the world's sixth mass extinction was ramping up. That my body was organized according to an anatomical archetype set forth in a world ruled by weird arthropods, constructed of an amazing mineralized material that went from being armor to an internal scaffold, forged by fish that dragged themselves onto land and refined by tens of millions of years in the trees. That as isolated as I might feel at any one point I'm still part of a bigger story of life on Earth. That tiny piece of my skeleton would be enough to show that the old pirate's adage is wrong. The dead really do tell tales.

ACKNOWLEDGMENTS

Much like skeletons themselves, books go through their own developmental processes. In this case, there were a few more growing pains than I expected. I took up the task of stretching beyond the realm of my usual expertise just when circumstance was about to deliver a series of unpleasant surprises. The fact that you are now holding this book at all is a credit to everyone who propped me up during each challenging stage of this manuscript's evolution.

Naturally, the editors and agents who've watched this project develop deserve immediate attention. Even though my previous agent Peter Tallack and the editor of my last book, Amanda Moon, did not join me on this project, I'm still indebted to them for their time and the result—intentional or not—which led me to realize that this was the book I needed to write. That set of changes led me to my indefatigable agent Deirdre Mullane—the most enthusiastic champion of writers I've ever met—and my insightful, ever-patient editor Courtney Young, who wisely kept nudging me to look beyond my fossiliferous comfort zone. Without Deirdre and Courtney, this book wouldn't exist.

Now here's the tricky part. In the process of writing any book, an author has far more people to thank than they are ever going to remember. There are friends, experts, and people who inadvertently provided important insights along the way. I have not yet mastered a way to remember them all and am entirely certain I am leaving someone out. (If you are expecting to find your name here and don't see it, I apologize.) But, so far as my recollection goes and in no particular order, I am thankful to Carl Zimmer, Deborah Blum, Mary Roach, Jennifer Ouellette, Maryn McKenna, Annalee Newitz, Mathew Wedel, Kevin Padian, John Hutchinson, Adam Huttenlocker, Sarah Werning, Michael Habib, John Hawks, Kristina Killgrove, Ted Daeschler, and more for their professional support, time, and insights as I've worked to spill my brains onto the page.

Of course, some of the people closest to me are those who know the least about what I'm working on. They see the stress and strain and competition of ideas, but the document itself is too precious to share until finished. This emotional support is as critical to completing a book as an instrument with which to write, and words truly fail to express how much I am indebted to Foxfeather Zenkova, Bethany Brookshire, Amber Hill, Alex Porpora, Carolyn Levitt-Bussian, Kit Robison, and Miriam Goldstein for their friendship and sympathetic ears. I hope you enjoy what I spent all that time fussing over.

Most of all, I'm forever grateful to my wife, Tracey, for her ceaseless support. She has never doubted me for a moment. Even at times when I've wanted to shut my laptop and leave it that way, she's urged me on. She believed in me when no one else did—when I didn't even have much more than a blog on WordPress and a penchant for writing—and every book I write is a symbol of her support. I may have fleshed things out, but the bones of my work belong to her.

A NOTE ABOUT THE ILLUSTRATIONS

The images used to front each chapter of *Skeleton Keys* come from, or were based upon, those in the 1733 book *Osteographia or the Anatomy of Bones*. It's one of the most striking bone manuals ever published.

Osteographia was written by English anatomist William Cheselden, now remembered as a pioneering surgeon. Despite Cheselden's academic notoriety, *Osteographia* is not quite like a medical textbook you'll find in a college bookstore. Modern books about skeletons and bone tissue are focused on accuracy and the multi-layered perspectives of bone's inherent nature, from the makeup of cells through the landmarks on a particular element. But Cheselden's book retains an aspect of the whimsical. There are detailed illustrations of skulls and individual elements, but the skeletons of *Osteographia* also come alive as if they didn't require any flesh at all. The human frames in Cheselden's book ponder, pose, despair, and pray, while additional plates—not included in this book as they relate to other species—show a hissing cat, sleeping dog, prancing antelope, and a bear seeming to ponder whether a tree hollow might have anything tasty in its upper branches, among others. This is what made *Osteographia* a fitting choice to draw illustrations from for *Skeleton Keys*. Cheselden's reconstructed skeletons are not static anatomical monuments, but they are a tribute to the liveliness and vitality of bones.

NOTES

INTRODUCTION: CUT TO THE BONE

1 **This happens to:** Mupparapu, Muralidhar, and Anitha Vuppalapati. "Ossification of Laryngeal Cartilages on Lateral Cephalometric Radiographs." *Angle Orthodontist* 75, no. 2 (2005): 196–201.

6 **Bone, in scientific:** Ota, Kinya, and Shigeru Kuratani. "Evolutionary Origin of Bone and Cartilage in Vertebrates." In *The Skeletal System*, edited by Olivier Pourquié, 1–18. Cold Spring Harbor, NY: Cold Spring Harbor Laboratory Press, 2009.

7 **"We determined to":** Steinbeck, John. *The Log from the Sea of Cortez.* New York: Penguin, 1941. p. 3.

ONE: FLESH OUT

17 **So Smithsonian taxidermist:** Ruane, Michael. "Natural History Museum Grants Professor's Dying Wish: A Display of His Skeleton." *Washington Post*, April 11, 2009. http://www.washingtonpost.com/wp-dyn/content/article/2009/04/10/AR2009041003357_3.html.

19 **That's why the:** Asma, Stephen. *Stuffed Animals and Pickled Heads*. New York: Oxford University Press, 2001. pp. 202–239.

21 **But as poetic:** Gould, Stephen. *Wonderful Life*. New York: W. W. Norton, 1989. pp. 71–75.

23 ***Pikaia* was one:** Walcott, Charles. *Cambrian Geology and Paleontology II: No. 5—Middle Cambrian Annelids*. Washington, DC: Smithsonian Institution, 1911.

24 **But when paleontologist:** Morris, Simon Conway, and Jean-Bernard Caron. "*Pikaia gracilens* Walcott, a Stem-Group Chordate from the Middle Cambrian of British Columbia." *Biological Reviews of the Cambridge Philosophical Society* 87, no. 2 (2012): 480–512.

26 **If you were:** "The Fossils." Royal Ontario Museum. http://burgess-shale.rom.on.ca/en/science/burgess-shale/03-fossils.php.

28 **Paleontologists who have:** Caron, Jean-Bernard, and Donald Jackson. "Paleoecology of the Greater Phyllopod Bed Community, Burgess Shale." *Palaeogeography, Palaeoclimatology, Palaeoecology* 258, no. 3 (2008): 222–56.

30 **Paleontologists thought that:** Morris, Simon Conway. "The Persistence of Burgess Shale–type Faunas: Implications for the Evolution of Deeper-Water Faunas." *Earth and Environmental Science Transactions of the Royal Society of Edinburgh* 80, no. 3–4 (1989): 271–83.

30 **In 2010, a:** Van Roy, Peter, Patrick J. Orr, Joseph P. Botting, Lucy A. Muir, Jakob Vinther, Bertrand Lefebvre, Khadija el Hariri, and Derek E. G. Briggs. "Ordovician Faunas of the Burgess Shale Type." *Nature* 465 (2010): 215–8.

TWO: BONES TO PICK

37 **Still, both Cope:** Editorial in the *American Naturalist*, June 1873, p. 384.

38 **Such procedures were:** Davidson, Jane. *The Bone Sharp*. Philadelphia: The Academy of Natural Sciences of Philadelphia, 1997.

39 **Dinosaurs and other:** "Hadrosaurus foulkii." Academy of Natural Sciences of Drexel University. http://ansp.org/exhibits/online-exhibits/stories/hadrosaurus-foulkii/.

39 **In fact, it:** Psihoyos, Louie, and John Knoebber. *Hunting Dinosaurs*. New York: Random House, 1994.

40 **Just like you:** Richter, Daniel, Rainer Grün, Renaud Joannes-Boyau, Teresa E. Steele, Fethi Amani, Mathieu Rué, Paul Fernandes et al. "The Age of the Hominin Fossils from Jebel Irhoud, Morocco, and the Origins of the Middle Stone Age." *Nature* 546, no. 7657 (2017): 293–6.

45 **Vertebrates missed the:** Satoh, Noriyuki, Daniel Rokhsar, and Teruaki Nishikawa. "Chordate Evolution and the Three-Phylum System." *Proceedings of the Royal Society B* 281, no. 1794 (2014); Erwin, Douglas, and Eric H. Davidson. "The Last Common Bilaterian Ancestor." *Development* 129, no. 13 (2002): 3021–32.

47 **The fossil fish:** Zhu, Min, Xiaobo Yu, Per Erik Ahlberg, Brian Choo, Jing Lu, Tuo Qiao, Qingming Qu et al. "A Silurian Placoderm with Osteichthyan-like Marginal Jaw Bones." *Nature* 502 (2013): 188–93.

48 ***Compagopiscis* lived 420:** Rücklin, Martin, Philip C. J. Donoghue, Zerina Johanson, Kate Trinajstic, Federica Marone, and Marco Stampanoni. "Development of Teeth and Jaws in the Earliest Jawed Vertebrates." *Nature* 491 (2012): 748–51.

50 **This similarity goes:** Zhu, Min, Wenjin Zhao, Liantao Jia, Jing Lu, Tuo Qiao, and Qingming Qu. "The Oldest Articulated Osteichthyan Reveals Mosaic Gnathostome Characters." *Nature* 458 (2009): 469–74.

50 **Somewhere along the:** Tanaka, Mikiko, Andrea Münsterberg, W. Gary Anderson, Alan R. Prescott, Neil Hazon, and Cheryll Tickle. "Fin Development in a Cartilaginous Fish and the Origin of Vertebrate Limbs." *Nature* 416 (2002): 527–31.

52 **The land had:** Wellman, Charles H., Peter L. Osterloff, and Uzma Mohiuddin. "Fragments of the Earliest Land Plants." *Nature* 425 (2003): 282–5.

53 **About 428 million:** Buatois, Luis, M. Gabriela Mangano, Jorge F. Genise, and Thomas N. Taylor. "The Ichnologic Record of the Continental Invertebrate Invasion; Evolutionary Trends in Environmental Expansion, Ecospace Utilization, and Behavioral Complexity." *PALAIOS* 13, no. 3 (1998): 217–40; Engel, Michael S., and David A. Grimaldi. "New Light Shed on the Oldest Insect." *Nature* 427 (2004): 627–30.

53 **Fish capable of:** MacIver, Malcolm A., Lars Schmitz, Ugurcan Mugan, Todd D. Murphey, and Curtis D. Mobley. "Massive Increase in Visual Range Preceded the Origin of Terrestrial Vertebrates." *PNAS* 114, no. 12 (2017): E2375–84.

55 **But at the:** Angielczyk, Ken, and Lars Schmitz. "Nocturnality in Synapsids Predates the Origin of Mammals by Over 100 Million Years." *Proceedings of the Royal Society B* 281, no. 1793 (2014).

61 **More than that:** Gebo, Daniel L. *Primate Comparative Anatomy*. Baltimore: Johns Hopkins University Press, 2014. pp. 8–10.

THREE: A BAG OF BONES

66 **"extensive swamps of bitumen":** Jefferson, George T. "People and the Brea: A Brief History of a Natural Resource." In *Rancho La Brea: Death Trap and Treasure Trove*, edited by John M. Harris, 3–8. Los Angeles: Natural History Museum of Los Angeles County, 2001.

66 **"the lake swallowed them":** Jefferson. "People and the Brea: A Brief History of a Natural Resource."

67 **Shallow puddles of:** Jefferson. "People and the Brea: A Brief History of a Natural Resource."

71 **British tabloids immediately:** McKenna, Josephine. "Embracing Figures at Pompeii 'Could Have Been Gay Lovers', After Scan Reveals They Are Both

Men." *Telegraph*, April 7, 2017. https://www.telegraph.co.uk/news/2017/04/07/embracing-figures-pompeii-could-have-been-gay-lovers-scan-reveals/.

72 **When associated or even:** Killgrove, Kristina. "Is That Skeleton Gay? The Problem with Projecting Modern Ideas onto the Past." *Forbes*, April 8, 2017. https://www.forbes.com/sites/kristinakillgrove/2017/04/08/is-that-skeleton-gay-the-problem-with-projecting-modern-ideas-onto-the-past/#48a373a830e7.

76 **Paleontologists have been:** Schweitzer, Mary Higby, Zhiyong Suo, Recep Avci, John M. Asara, Mark A. Allen, Fernando Teran Arce, and John R. Horner. "Analyses of Soft Tissue from *Tyrannosaurus rex* Suggest the Presence of Protein." *Science* 316, no. 5822 (2007): 277–80.

77 **Then, with the:** Clarke, B. "Normal Bone Anatomy and Physiology." *CJASN* 3, supplement 3 (2008): S131–9.

77 **All walled up, the:** Buenzli, Pascal R., and Natalie A. Sims. "Quantifying the Osteocyte Network in the Human Skeleton." *Bone* 75 (2015): 144–50.

82 **The two types:** Alexander, R. McNeill. *Bones: The Unity of Form and Function.* New York: Macmillan, 1994. pp. 24–57.

83 **In addition to:** Molnár, M., I. János, L. Szűcs, and L. Szathmáry. "Artificially Deformed Crania from the Hun-Germanic Period (5th–6th century AD) in Northeastern Hungary: Historical and Morphological Analysis." *Neurosurgical Focus* 36, no. 4 (2014): E1; Clark, J. "The Distribution and Cultural Context of Artificial Cranial Modification in the Central and Southern Philippines." *Asian Perspectives* 52, no. 1 (2013): 28–42; Durband, Arthur C. "Brief Communication: Artificial Cranial Modification in Kow Swamp and Cohuna." *American Journal of Physical Anthropology* 155, no. 1 (2014): 173–8; Gerszten, Peter C. "An Investigation into the Practice of Cranial Deformation Among the Pre-Columbian Peoples of Northern Chile." *International Journal of Osteoarchaeology* 3, no. 2 (1993): 87–98.

84 **In southern France:** O'Brien, Tyler G., Lauren R. Peters, and Marc E. Hines. "Artificial Cranial Deformation: Potential Implications for Affected Brain Function." *Anthropology* 1, no. 3 (2013): 1–6.

85 **Some of these:** O'Brien, T., and A. M. Stanley. "Boards and Cords: Discriminating Types of Artificial Cranial Deformation in Prehispanic South Central Andean Populations." *International Journal of Osteoarchaeology* 23, no. 4 (2013): 459–70.

85 **The simplest explanation:** Okumura, Mercedes. "Differences in Types of Artificial Cranial Deformation Are Related to Differences in Frequencies of Cranial and Oral Health Markers in Pre-Columbian Skulls from Peru." *Boletim do Museu Paraense Emílio Goeldi: Ciências Humanas* 9, no. 1 (2014): 15–26.

87 **In fact, in:** Killgrove, Kristina. "Here's How Corsets Deformed the Skeletons of Victorian Women." *Forbes*, November 16, 2015. https://www.forbes.com/sites/kristinakillgrove/2015/11/16/how-corsets-deformed-the-skeletons-of-victorian-women/#45121f7b799c; Gibson, Rebecca. "Effects of Long

Term Corseting on the Female Skeleton: A Preliminary Morphological Examination." *Nexus* 23, no. 2 (2015): 45–60.

FOUR: BONE SHAKING

96 **Fossils such as *Ardipithecus*:** Lovejoy, C. Owen, Gen Suwa, Linda Spurlock, Berhane Asfaw, and Tim D. White. "The Pelvis and Femur of *Ardipithecus ramidus*: The Emergence of Upright Walking." *Science* 326, no. 5949 (2009): 71e1–6.

96 **Part of that:** Aiello, Leslie, and Christopher Dean. *An Introduction to Human Evolutionary Anatomy*. London: Elsevier, 2002. p. 285.

106 **Two complementary studies:** Ryan, Timothy M., and Colin N. Shaw. "Gracility of the Modern *Homo sapiens* Skeleton Is the Result of Decreased Biomechanical Loading." *PNAS* 112, no 2. (2015): 372–7; Chirchir, Habiba, Tracy L. Kivell, Christopher B. Ruff, Jean-Jacques Hublin, Kristian J. Carlson, Bernhard Zipfel, and Brian G. Richmond. "Recent Origin of Low Trabecular Bone Density in Modern Humans." *PNAS* 112, no. 2 (2015): 366–71.

108 **Space stations in:** "Space Bones." NASA Science, October 1, 2001. https://science.nasa.gov/science-news/science-at-nasa/2001/ast01oct_1/.

109 **In a 2015:** McGee-Lawrence, Meghan, Patricia Buckendahl, Caren Carpenter, Kim Henriksen, Michael Vaughan, and Seth Donahue. "Suppressed Bone Remodeling in Black Bears Conserves Energy and Bone Mass During Hibernation." *Journal of Experimental Biology* 218, pt. 13 (2015): 2067–74.

FIVE: STICKS AND STONES

116 **The reptile suffered:** Reisz, Robert R., Diane M. Scott, Bruce R. Pynn, and Sean P. Modesto. "Osteomyelitis in a Paleozoic Reptile: Ancient Evidence for Bacterial Infection and Its Evolutionary Significance." *Naturwissenschaften* 98, no. 6 (2011): 551–5.

117 **Sometimes an open:** Anné, Jennifer, Brandon P. Hedrick, and Jason P. Schein. "First Diagnosis of Septic Arthritis in a Dinosaur." *Royal Society Open Science* 3, no. 8 (2016): 160222.

118 **While tooth cavities:** Humphrey, Louise T., Isabelle De Groote, Jacob Morales, Nick Barton, Simon Collcutt, Christopher Bronk Ramsey, and Abdeljalil Bouzouggar. "Earliest Evidence for Caries and Exploitation of Starchy Plant Foods in Pleistocene Hunter-Gatherers from Morocco." *PNAS* 111, no. 3 (2014): 954–9.

118 **And, knowing what:** Sheridan, Kerry. "Eating Nuts Caused Tooth Decay in Hunter-Gatherers." PhysOrg, January 6, 2014. https://phys.org/news/2014-01-nuts-tooth-hunter-gatherers.html; Oxilia, Gregorio, Marco Peresani, Matteo Romandini, Chiara Matteucci, Cynthianne Debono Spiteri, Amanda G.

Henry, Dieter Schulz et al. "Earliest Evidence of Dental Caries Manipulation in the Late Upper Palaeolithic." *Scientific Reports* 5 (2015): 12150.

118 **KNM-ER 1808 is:** "KNM-ER 1808." Smithsonian National Museum of Natural History, March 30, 2016. http://humanorigins.si.edu/evidence/human-fossils/fossils/knm-er-1808.

118 **Lesions marked her:** Dolan, Sean Gregory. "A Critical Examination of the Bone Pathology on KNM-ER 1808, a 1.6 Million Year Old *Homo erectus* from Koobi Fora, Kenya." Master's thesis, New Mexico State University, 2011.

119 **As you might:** Walker, Alan, M. R. Zimmerman, and R. E. F. Leakey. "A Possible Case of Hypervitaminosis A in *Homo erectus*." *Nature* 296 (1982): 248–50; Skinner, Mark. "Bee Brood Consumption: An Alternative Explanation for Hypervitaminosis A in KNM-ER 1808 (*Homo erectus*) from Koobi Fora, Kenya." *Journal of Human Evolution* 20, no. 6 (1991): 493–503.

121 **But the physiological:** "Persistence of Epiphyseal Line in the Iliac Crest," *Forbes*. https://www.forbes.com/pictures/gked45glfl/persistence-of-epiphysea/#273d86f71da6.

121 **The same might:** Killgrove, Kristina. "Skeletons of Two Possible Eunuchs Discovered in Ancient Egypt." *Forbes*, April 28, 2017. https://www.forbes.com/sites/kristinakillgrove/2017/04/28/skeletons-of-two-possible-eunuchs-discovered-in-ancient-egypt/#3d82d8251f55.

123 **As orthopedic doctor:** Kaplan, Frederick. "The Skeleton in the Closet." *Gene* 528, no. 1 (2013): 7–11.

123 **In cases of:** Ibid.; De La Hoz Polo, Marcela, Monica Khanna, and Miny Walker. "Young Woman Who Presents with Shortness of Breath." *Skeletal Radiology* 46, no. 1 (2017): 143–5.

123 **"They likewise arise":** Kaplan, Frederick, Martine Le Merrer, David L. Glaser, Robert J. Pignolo, Robert Goldsby, Joseph A. Kitterman, Jay Groppe, and Eileen M. Shore. "Fibrodysplasia Ossificans Progressiva." *Best Practice & Research Clinical Rheumatology* 22, no. 1 (2008): 191–205.

124 **Even centuries after:** Kamal, Achmad Fauzi, Robin Novriansyah, Rahyussalim, Yogi Prabowo, and Nurjati Chairani Siregar. "Fibrodysplasia Ossificans Progressiva: Difficulty in Diagnosis and Management: A Case Report and Literature Review." *Journal of Orthopaedic Case Reports* 5, no. 1 (2015): 26–30.

124 **And humans aren't:** Warren, H. B., and J. L. Carpenter. "Fibrodysplasia Ossificans in Three Cats." *Veterinary Pathology* 21, no. 5 (1984): 495–9; Guilliard, M. J. "Fibrodysplasia Ossificans in a German Shepherd Dog." *Journal of Small Animal Practice* 42, no. 11 (2001): 550–3.

125 **He was almost:** Angier, Natalie. "Bone, a Masterpiece of Elastic Strength." *New York Times*, April 27, 2009. https://www.nytimes.com/2009/04/28/science/28angi.html.

125 **To date, there's:** Kaplan, "The Skeleton in the Closet."

125 **Kaplan says of the research:** Ibid.

126 **So that's why:** Yilmaz, Ibrahim Edhem, Yagil Barazani, and Basir Tareen. "Penile Ossification: A Traumatic Event or Evolutionary Throwback? Case Report and Review of the Literature." *Canadian Urological Association Journal* 7, no. 1–2 (2013): E112–4.

127 **The body lay:** "Evidence of Trepanation Found in 7,000 Year Old Skull from Sudan," *Archaeology News Network*, July 1, 2016. https://archaeologynewsnetwork.blogspot.com/2016/07/evidence-of-trepanation-found-in-7000.html#FRWpjhgGy0oZP9Mx.97.

127 **More than eight hundred examples:** Watson, Traci. "Amazing Things We've Learned From 800 Ancient Skull Surgeries." *National Geographic*, June 30, 2016. http://news.nationalgeographic.com/2016/06/what-is-trepanation-skull-surgery-peru-inca-archaeology-science/.

128 **The number of:** Andrushko, Valerie A., and John W. Verano. "Prehistoric Trepanation in the Cuzco Region of Peru: A View into an Ancient Andean Practice." *American Journal of Physical Anthropology* 137, no. 1 (2008): 4–13.

128 **"Trepanations were placed":** Ibid.

129 **In 2015, Jenna:** Wade, Lizzie. "Skeletons from Hospital Graveyard Shed Light on Early Dissections." *Science*, February 15, 2015. http://www.sciencemag.org/news/2015/02/skeletons-hospital-graveyard-shed-light-early-dissections?rss=1.

130 **In 2017, archaeologist:** Smith-Guzmán, Nicole, Jeffrey A. Toretsky, Jason Tsai, and Richard G. Cooke. "A Probable Primary Malignant Bone Tumor in a Pre-Columbian Human Humerus from Cerro Brujo, Bocas del Toro, Panamá." *International Journal of Paleopathology*, 2017.

SIX: THE NEARER THE BONE, THE SWEETER THE MEAT

136 **Anatomists have compared:** Day, Michael H., and Robert W. Pitcher-Wilmott. "Sexual Differentiation in the Innominate Bone Studied by Multivariate Analysis." *Annals of Human Biology* 2, no. 2 (1975): 143–51.

137 **The reason for:** Cunningham, C., L. Scheuer, and S. Black. *Developmental Juvenile Osteology*. London: Academic Press, 2016. p. 16.

138 **We now know:** Hoffman, D. L., C. D. Standish, M. Garcia-Diez, P. B. Pettitt, J. A. Milton, J. Zilhão, J. J. Alcolea-Gonzalez et al. "U-Th Dating of Carbonate Crusts Reveals Neandertal Origin of Iberian Cave Art." *Science* 359, no. 6378 (2018): 912–5.

139 **For decades, archaeologists:** Paul B. Pettitt. "The Neanderthal Dead: Exploring Mortuary Variability in Middle Palaeolithic Eurasia." *Before Farming* 1 (2002): 1–26.

139 **"It is difficult":** d'Errico, Francesco, Christopher Henshilwood, Graeme Lawson, Marian Vanhaeren, Anne-Marie Tillier, Marie Soressi, Frédérique

Bresson et al. "Archaeological Evidence for the Emergence of Language, Symbolism, and Music—An Alternative Multidisciplinary Perspective." *Journal of World Prehistory* 17, no. 1 (2003): 1–70.

139 **From Uzbekistan to:** Langley, Michelle C., Christopher Clarkson, and Sean Ulm. "Behavioural Complexity in Eurasian Neanderthal Populations: A Chronological Examination of the Archaeological Evidence." *Cambridge Archaeological Journal* 18, no. 3 (2008): 289–307.

142 **They were intentionally:** Gresky, Julia, Juliane Haelm, and Lee Clare. "Modified Human Crania from Göbekli Tepe Provide Evidence for a New Form of Neolithic Skull Cult." *Science Advances* 3, no. 6 (2017): e1700564.

143 **Part of their:** Manseau, Peter. *Rag and Bone: A Journey Among the World's Holy Dead.* New York: Henry Holt and Company, 2009. p. 7

144 **This terrifying legacy:** Quenneville, Guy. "'We Love Him to Bits': Severed Arm of St. Francis Xavier Draws Hundreds in Saskatoon." CBC, January 18, 2018. http://www.cbc.ca/news/canada/saskatoon /love-him-bits-severed-arm-st-francis-xavier-display-saskatoon-1.4493026.

144 **"I've photographed at":** Paulas, Rick. "The Weird and Fraudulent World of Catholic Relics." *Vice*, March 4, 2015. https://www.vice.com/en_us/article /jmbwzg/the-weird-and-fraudulent-world-of-catholic-relics-456.

145 **You can bid:** eBay, s.v. "Elvis Hair." Accessed April 27, 2018. http://www.ebay .com/bhp/elvis-hair; "Marilyn Monroe Hair." Paul Fraser Collectibles. Accessed April 27, 2018. https://store.paulfrasercollectibles.com/products /marilyn-monroe-authentic-strand-of-hair.

145 **That makes the:** Bello, Silvia M., Simon A. Parfitt, and Chris B. Stringer. "Earliest Directly-Dated Human Skull-Cups." *PLOS One* 6, no. 2 (2011): e17026.

145 **"Cut-marks on the areas":** Ibid.

147 **But a reinterpretation:** Bonogofsky, M. "Cranial Modeling and Neolithic Bone Modification at 'Ain Ghazal: New Interpretations." *Paléorient* 27, no. 2 (2001): 141–6.

147 **Another, from a site:** Bonogofsky, "Cranial Modeling and Neolithic Bone Modification at 'Ain Ghazal."

148 **But it's still:** Goren, Yuval, A. Nigel Goring-Morris, and Irena Segal. "The Technology of Skull Modelling in the Pre-Pottery Neolithic B (PPNB): Regional Variability, the Relation of Technology and Iconography and Their Archaeological Implications." *Journal of Archaeological Science* 28, no. 7 (2001): 671–90.

148 **The artifact looks:** Moore, Ken. "Instruments of Macabre Origin." The Met, July 7, 2014. http://www.metmuseum.org/blogs/of-note/2014/skull-lyre.

149 **"When visitors come":** Larson, Frances. *Severed: A History of Heads Lost and Heads Found.* New York: Liveright, 2014. pp. 17–24.

150 **"To be gnawed out":** Dickey, Colin. *Cranioklepty.* Lakewood, CO: Unbridled Books, 2009. p. 16.

SEVEN: BAD TO THE BONE

154 **Writing for *Forbes*:** Killgrove, Kristina. "A Summer Day in the Life of a Roman Bioarchaeologist." *Forbes*, July 27, 2017. https://www.forbes.com/sites/kristinakillgrove/2017/07/27/a-summer-day-in-the-life-of-a-roman-bioarchaeologist/#2b5ab59f4cb6.

155 **"We measure them":** Ibid.

155 **Upon examination, the:** Pearson, Michael. "Uprooted Tree Reveals a Violent Death from 1,000 Years Ago." CNN, September 15, 2015. https://www.cnn.com/2015/09/15/europe/ireland-tree-skeleton-discovery-feat/index.html.

160 **"Owing to the lack":** Buckley, Richard, Mathew Morris, Jo Appleby, Turi King, Deirdre O'Sullivan, and Lin Foxhall. "'The King in the Car Park': New Light on the Death and Burial of Richard III in the Grey Friars Church, Leicester, in 1485." *Antiquity* 87 (2013): 519–38.

160 **"Archaeologists today do not":** Ibid.

161 **The prone skeleton:** Ibid.

161 **"The body appears to have":** Ibid.

162 **The maximum window:** King, Turi, Gloria Gonzalez Fortes, Patricia Balaresque, Mark G. Thomas, David Balding, Pierpaolo Maisano Delser, Rita Neumann et al. "Identification of the Remains of King Richard III." *Nature Communications* 5 (2014): 1–8.

163 **Even though Richard III:** Ibid.

165 **We're animals, too:** Lamb, Angela L., Jane E. Evans, Richard Buckley, and Jo Appleby. "Multi-isotope Analysis Demonstrates Significant Lifestyle Changes in King Richard III." *Journal of Archaeological Science* 50 (2014): 559–65.

167 **What they found:** Appleby, Jo, Piers D. Mitchell, Claire Robinson, Alison Brough, Guy Rutty, Russell A. Harris, David Thompson, and Bruno Morgan. "The Scoliosis of Richard III, Last Plantagenet King of England: Diagnosis and Clinical Significance." *Lancet* 383, no. 9932 (2014): 1944.

168 **Appleby and colleagues write:** Ibid.

168 **"A good tailor and custom-made":** Ibid.

169 **And working from:** Lewis, Jason. "Identifying Sword Marks on Bone: Criteria for Distinguishing Between Cut Marks Made by Different Classes of Bladed Weapons." *Journal of Archaeological Science* 35, no. 7 (2008): 2001–8.

169 **You can still see the:** Appleby, Jo, Guy N. Rutty, Sarah V. Hainsworth, Robert C. Woosnam-Savage, Bruno Morgan, Alison Brough, Richard W. Earp et al. "Perimortem Trauma in King Richard III: A Skeletal Analysis." *Lancet* 385, no. 9964 (2015): 253–9.

170 **"This injury was associated . . . from above":** Ibid.

170 **The damage pattern suggests . . . pointing downwards":** Ibid.
171 **"Reconstruction of the pelvis . . . life-threatening":** Ibid.

EIGHT: BONES OF CONTENTION

178 **In the realm:** Stemmler, Joan K. "The Physiognomical Portraits of Johann Caspar Lavater." *Art Bulletin* 75, no. 1 (1993): 151–68.
178 **Lavater's lavishly illustrated:** Percival, Melissa. "Johann Caspar Lavater: Physiognomy and Connoisseurship." *Journal for Eighteenth-Century Studies* 26, no. 1 (2003): 77–90.
179 **"The pronouncements of":** van Whye, John. "Was Phrenology a Reform Science? Towards a New Generalization for Phrenology." *History of Science* 42, no. 137, pt. 3 (2004): 313–31.
180 **The cause of:** Rafter, Nicole. "The Murderous Dutch Fiddler: Criminology, History and the Problem of Phrenology." *Theoretical Criminology* 9, no. 1 (2005): 65–96.
180 **If about to:** van Whye, "Was Phrenology a Reform Science?"
181 **In Australia, historian:** McGregor, Russell. *Imagined Destinies: Aboriginal Australians and the Doomed Race Theory, 1880–1939.* Victoria, Australia: Melbourne University Press, 1997.
182 **Phrenology was used:** Bank, Andrew. "Of 'Native Skulls' and 'Noble Caucasians': Phrenology in Colonial South Africa." *Journal of Southern African Studies* 22, no. 3 (1996): 387–403.
182 **European soldiers made:** Webb, Denver A. "War, Racism, and the Taking of Heads: Revisiting Military Conflict in the Cape Colony and Western Xhosaland in the Nineteenth Century." *Journal of African History* 56, no. 1 (2015): 37–55.
183 **The inspiration for:** Renschler, Emily S., and Janet Monge. "The Samuel George Morton Cranial Collection." *Expedition* 50, no. 3 (2008): 30–8.
186 **"About thirty years":** "To Henry Fawcett 18 September [1861]," Darwin Correspondence Project, University of Cambridge. https://www.darwinpro ject.ac.uk/letter/DCP-LETT-3257.xml.
187 **"If this doctrine be unfounded":** Gould, Stephen Jay. *The Mismeasure of Man* (New York: W.W. Norton and Company, 1996): 84.
188 **But as historian:** Stanton, William. *The Leopard's Spots.* Chicago: University of Chicago Press, 1960.
188 **In a letter:** Ibid., 137.
189 **he might have:** Weisberg, Michael, and Diane B. Paul. "Morton, Gould, and Bias: A Comment on 'The Mismeasure of Science." *PLOS Biology* 14, no. 4 (2016): e1002444.
189 **In 2011, a:** Lewis, Jason E., David DeGusta, Marc R. Meyer, Janet M. Monge, Alan E. Mann, and Ralph L. Holloway. "The Mismeasure of Science: Stephen

Jay Gould Versus Samuel George Morton on Skulls and Bias." *PLOS Biology* 9, no. 6 (2011): e1001071.

190 **"It is hard to see how":** Kaplan, Jonathan, Massimo Pigliucci, and Joshua Alexander Banta. "Gould on Morton, Redux: What can the debate reveal about the limits of data?" *Studies in History of Philosophy of Biological and Biomedical Sciences* 52 (2015): 22–31.

192 **Head form which has always been:** Thomas, David Hurt. *Kennewick Man*. New York: Basic Books, 2000. p. 104.

192 **It was only:** Marks, Jonathan. "The Two 20th-Century Crises of Racial Anthropology." In *Histories of American Physical Anthropology in the 20th Century*, edited by Michael A. Little and Kenneth A. R. Kennedy, Lanham, MD: Lexington Books, 2010. pp. 187–206.

193 **This was difficult work:** Ibid.

194 **The three factors:** Redman, Samuel J. *Bone Rooms*. Cambridge, MA: Harvard University Press, 2016. p. 222.

194 **Anthropologist Kenneth Kennedy:** Kennedy, Kenneth A. R. "Principal Figures in Early 20th-Century Physical Anthropology: With Special Treatment of Forensic Anthropology." In Little and Kennedy, *Histories of American Physical Anthropology*, 105–26.

195 **Those efforts forced:** Hemmer, Nicole. "'Scientific Racism' Is on the Rise on the Right. But It's Been Lurking There for Years." *Vox*, March 28, 2017. https://www.vox.com/the-big-idea/2017/3/28/15078400/scientific-racism-murray-alt-right-black-muslim-culture-trump.

195 **In early 2018:** Devlin, Hannah. "First Modern Britons Had 'Dark to Black' Skin, Cheddar Man DNA Analysis Reveals." *Guardian*, February 7, 2018. https://www.theguardian.com/science/2018/feb/07/first-modern-britons-dark-black-skin-cheddar-man-dna-analysis-reveals.

NINE: SKELETONS IN THE CLOSET

199 **Early on the chilly morning:** Williams, Weston. "Burial of 9,000-Year-Old Kennewick Man Lays to Rest a 20-Year-Old Debate." *Christian Science Monitor*, February 21, 2017. http://www.csmonitor.com/Science/2017/0221/Burial-of-9-000-year-old-Kennewick-Man-lays-to-rest-a-20-year-old-debate.

200 **The skeleton scientists:** Owsley, Douglas W., and Richard L. Jantz. *Kennewick Man*. College Station, TX: Texas A&M University Press, 2014.

200 **Everything changed on:** Ibid.

202 **Almost thirty years on:** Ousley, Stephen D., William T. Billeck, and R. Eric Hollinger. "Federal Repatriation Legislation and the Role of Physical Anthropology in Repatriation." *Yearbook of Physical Anthropology* 48 (2005): 2–32.

203 **The Kennewick case:** Bruning, Susan B. "Complex Legal Legacies: The Native American Graves Protection and Repatriation Act, Scientific Study, and Kennewick Man." *American Antiquity* 71, no. 3 (2006): 501–21.

205 **But it's difficult to:** Watkins, Joe. "Becoming American or Becoming Indian?" *Journal of Social Archaeology* 4, no. 1 (2004): 60–80.

206 **"The conflict between Indians":** Maureen Konkle. *Writing Indian Nations.* Chapel Hill: University of North Carolina Press, 2004. p. 292.

206 **In a paper:** Sauer, Norman. "Forensic Anthropology and the Concept of Race: If Races Don't Exist, Why Are Forensic Anthropologists So Good at Identifying Them?" *Social Science & Medicine* 34, no. 2 (1992): 107–11.

207 **As Pawnee historian:** Echo-Hawk, Roger, and Larry J. Zimmerman. "Beyond Racism: Some Opinions About Racialism and American Archaeology." *American Indian Quarterly* 30, no. 3/4 (2006): 461–85.

208 **In 1971, construction:** Killgrove, Kristina. "How One Anthropologist Balances Human Skeletons and Human Rights." *Forbes*, March 17, 2017. https://www.forbes.com/sites/kristinakillgrove/2017/03/17/how-one-anthropologist-balances-human-skeletons-and-human-rights/#679ad57c2a1f.

209 **"I remember reading":** Lippert, Dorothy. "Repatriation and the Initial Steps Taken on Common Ground." *SAA Archaeological Record* 15, no. 1 (2015): 36–8.

210 **What they found:** Rasmussen, Morten, Martin Sikora, Anders Albrechtsen, Thorfinn Sand Korneliussen, J. Víctor Moreno-Mayar, G. David Poznik, Christoph P. E. Zollikofer et al. "The Ancestry and Affiliations of Kennewick Man." *Nature* 523 (2015): 455–8.

211 **He continued to:** Doughton, Sandi. "What's Next for Kennewick Man, Now That DNA Says He's Native American?" *Seattle Times*, June 18, 2015. http://www.seattletimes.com/seattle-news/science/kennewick-man-mystery-solved-dna-says-hes-native-american/.

211 **"Because of the":** Ibid.

212 **They buried the:** Rosenbaum, Cary. "Ancient One, Also Known as Kennewick Man, Repatriated." *Tribal Tribune*, February 18, 2017. http://www.tribaltribune.com/news/article_aa38c0c2-f66f-11e6-9b50-7bb1418f3d3d.html.

212 **In 1989, years:** Rose, Jerome C., Thomas J. Green, and Victoria D. Green. "NAGPRA Is Forever: Osteology and the Repatriation of Skeletons." *Annual Review of Anthropology* 25 (1996): 81–103.

213 **The new model:** Colwell, Chip, and Stephen E. Nash. "Repatriating Human Remains in the Absence of Consent." *The SAA Archaeological Record* 15, no. 1 (2015): 14–6.

215 **During the late:** Greenfieldboyce, Nell. "The Saga of the Irish Giant's Bones Dismays Medical Ethicists." NPR, March 13, 2017. http://www.npr.org/sections

/health-shots/2017/03/13/514117230/the-saga-of-the-irish-giants-bones -dismays-medical-ethicists.

215 **When, in 2011:** Doyal, Len, and Thomas Muinzer. "Should the Skeleton of 'the Irish giant' Be Buried at Sea?" *BMJ* 343 (2011): d7597.

216 **Yet, despite von Hagens's:** Ulaby, Neda. "Origins of Exhibited Cadavers Questioned." NPR, August 11, 2006. https://www.npr.org/templates/story /story.php?storyId=5637687; Harding, Luke. "Von Hagens Forced to Return Controversial Corpses to China. *Guardian*, January 23, 2004. https://www .theguardian.com/world/2004/jan/23/arts.china.

216 **Competing exhibitions—such:** Perkel, Colin. "'Bodies Revealed' Exhibit May Be Using Executed Chinese Prisoners, Says Rights Group." CBC, September 6, 2014. http://www.cbc.ca/news/canada/bodies-revealed-exhibit-may-be-using -executed-chinese-prisoners-says-rights-group-1.2757908.

217 **This is what:** Carney, Scott. *The Red Market*. New York: William Morrow, 2011.

217 **In 2007, for:** Carney, Scott. "Into the Heart of India's Underground Bone Trade." NPR, November 29, 2007. https://www.npr.org/templates/story/story .php?storyId=16678816.

217 **The company had been buying:** Carney, Scott. "Inside India's Underground Trade in Human Remains." *Wired*, November 27, 2007. https://www.wired .com/2007/11/ff-bones/.

217 **There seemed to:** "Young Brothers." India Mart. https://www.indiamart.com /youngbrothers/profile.html.

218 **So the trade:** Andrabi, Jalees. "Ban Fails to Stop Sales of Human Bones." *National*, February 13, 2009. https://www.thenational.ae/world/asia /ban-fails-to-stop-sales-of-human-bones-1.528471.

218 **Bones from people:** Carney, Scott. "Inside India's Underground Trade in Human Remains." *Wired*, November 27, 2007. https://www.wired.com/2007 /11/ff-bones/.

218 **Their teachers insist:** Cohen, Margot. "Booming Business in Bones: Demand for Real Human Skeletons Surges in India." *National*, December 28, 2015. https:// www.thenational.ae/world/booming-business-in-bones-demand-for-real-human -skeletons-surges-in-india-1.111275.

219 **The market even:** Suri, Manveena. "India: Police Arrest 8 in Human Bone Smuggling Ring." CNN, March 23, 2017. https://www.cnn.com/2017/03/23 /asia/india-bone-smuggling/index.html.

219 **People who need:** Kiser, Margot. "Burundi's Black Market Skull Trade." *Daily Beast*, January 26, 2014. https://www.thedailybeast.com /burundis-black-market-skull-trade.

219 **The online store:** "Real Human Bones for Sale," Bone Room. https://www .boneroom.com/store/c44/Human_Bones.html.

220 **Private collector Ryan:** Davis, Simon. “Meet the Living People Who Collect Dead Human Remains.” *Vice*, July 13, 2015. https://www.vice.com/en_us/article/wd7jd5/meet-the-living-people-who-collect-human-remains-713.

220 **Artist Zane Wylie:** “Real Human Skulls,” Replica and Real Human Skull Props by Zane Wylie, accessed June 12, 2018, https://realhumanskull.com/t/real-human-skull.

220 **The craft-centric Etsy:** Marsh, Tanya D. “Rethinking Laws Permitting the Sales of Human Remains.” *Huffington Post*, August 13, 2012. https://www.huffingtonpost.com/tanya-d-marsh/laws-permitting-human-remains_b_1769082.html.

220 **In this case:** Hugo, Kristin. “Human Skulls Are Being Sold Online, but Is It Legal?” *National Geographic*, August 23, 2016. https://news.nationalgeographic.com/2016/08/human-skulls-sale-legal-ebay-forensics-science/.

220 **Christine Halling and:** Halling, Christine L., and Ryan M. Seidemann. “They Sell Skulls Online?! A Review of Internet Sales of Human Skulls on eBay and the Laws in Place to Restrict Sales.” *Journal of Forensic Sciences* 61, no. 5 (2016): 1322–6.

220 **This came as:** Seidemann, Ryan M., Christopher M. Stojanowski, and Frederick J. Rich. “The Identification of a Human Skull Recovered from an eBay Sale.” *Journal of Forensic Sciences* 54, no. 6 (2009): 1247–53.

221 **Archaeologists such as Damien:** Killgrove, Kristina. “This Archaeologist Uses Instagram to Track the Human Skeleton Trade.” *Forbes*, July 6, 2016. https://www.forbes.com/forbes/welcome/?toURL=https://www.forbes.com/sites/kristinakillgrove/2016/07/06/this-archaeologist-uses-instagram-to-track-the-human-skeleton-trade/&refURL=&referrer=#267c05756598; Huffer, Damien, and Shawn Graham. “The Insta-Dead: The Rhetoric of the Human Remains Trade on Instagram.” *Internet Archaeology* 45 (2017). doi: 10.11141/ia.45.5.

221 **“In the same”:** Huffer and Graham, “The Insta-Dead.”

221 **“The ability to”:** Ibid.

222 **“One particular purveyor”:** Hugo, “Human Skulls Are Being Sold Online.”

TEN: BONE DEEP

231 **Modern hyenas had:** Rudwick, Martin. *Bursting the Limits of Time*. Chicago: University of Chicago Press, 2005. pp. 623–38.

234 **All of this:** Weigelt, Johannes. *Recent Vertebrate Carcasses and Their Paleobiological Implications.* Chicago: University of Chicago Press, 1989.

235 **In fact, DNA:** Allentoft, Morten E., Matthew Collins, David Harker, James Haile, Charlotte L. Oskam, Marie L. Hale, Paula F. Campos et al. “The Half-Life of DNA in Bone: Measuring Decay Kinetics in 158 Dated Fossils.” *Proceedings of the Royal Society B* 279, no. 1748 (2012): 4724–33.

236 **Such was the:** Gu, Jun-Jie, Fernando Montealegre-Z, Daniel Robert, Michael S. Engel, Ge-Xia Qiao, and Dong Ren. "Wing Stridulation in a Jurassic Katydid (Insecta, Orthoptera) Produced Low-Pitched Musical Calls to Attract Females." *PNAS* 109, no. 10 (2012): 3868–73.

237 **Some of the 1.8-million-year-old:** Brochu, Christopher A., Jackson Njau, Robert J. Blumenschine, and Llewellyn D. Densmore. "A New Horned Crocodile from the Plio-Pleistocene Hominid Sites at Olduvai Gorge, Tanzania." *PLOS One* 5, no. 2 (2010): e9333.

237 **Another fossil from:** Brain, C. K. "An Attempt to Reconstruct the Behaviour of Australopithecines: The Evidence for Interpersonal Violence." *Zoologica Africana* 7, no. 1 (1972): 379–401.

238 **And the famous:** Boaz, Noel T., Russell L. Ciochon, Qinqi Xu, and Jinyi Liu. "Mapping and Taphonomic Analysis of the *Homo erectus* Loci at Locality 1 Zhoukoudian, China." *Journal of Human Evolution* 46, no. 5 (2004): 519–49.

239 **For a few:** Bottjer, David J., Walter Etter, James W. Hagadorn, and Carol M. Tang. *Exceptional Fossil Preservation.* New York: Columbia University Press, 2002.

241 **Still, if you:** Sansom, Robert S., Sarah E. Gabbott, and Mark A. Purnell. "Atlas of Vertebrate Decay: A Visual and Taphonomic Guide to Fossil Interpretation." *Palaeontology* 56, no. 3 (2013): 457–74.

INDEX